HF Amateur Radio

Ian Poole, G3YWX

Radio Society of Great Britain

Published by the Radio Society of Great Britain, Cranborne Road, Potters Bar, Herts EN6 3JE.

First published 2001

© Radio Society of Great Britain, 2001. All rights reserved. No part of this publication may be reproduced, stored in a retrieval system, or transmitted, in any form or by any means, electronic, mechanical, photocopying, recording or otherwise, without the prior written permission of the Radio Society of Great Britain.

ISBN 1 872309 75 5

Publisher's note
The opinions expressed in this book are those of the author and not necessarily those of the RSGB. While the information presented is believed to be correct, the author, the publisher and their agents cannot accept responsibility for consequences arising from any inaccuracies or omissions.

Cover design: Bob Ryan.
Cover filmsetting: JJays, Southend.
Illustrations: Bob Ryan.
Typography: Ray Eckersley, Seven Stars Publishing, Marlow.
Production: Mike Dennison.

Printed in Great Britain by Black Bear Press, Cambridge.

Contents

Preface v
1 An introduction to the HF bands 1
2 Radio wave propagation 7
3 Types of transmission 19
4 Receivers 33
5 Transmitters 49
6 Antennas 63
7 Bands and band plans 81
8 On the bands 91
9 Setting up the radio station 105
   Appendix: Abbreviations and codes 115
   Index 119

Preface

THE HF (short-wave) bands have always been one of my favourite areas of amateur radio. I can still remember the excitement of making my first contact on 80m, then my first contact outside the UK, and later, early one morning, making contacts on 20m with the west coast of the USA. The fascination of being able to make contacts over these vast distances without the use of any additional electronics beyond those in the two stations involved still gives me a sense of awe and excitement.

Hopefully this book will help others, whether new to the hobby or somewhat more experienced, to gain more from amateur radio, and to experience some of the enjoyment and excitement I have felt over the years. Despite all the new technology that is available today, HF band operating still has a lot to offer, both in terms of the enjoyment of operating and talking to people all around the globe as well as exploring the limits of science and technology.

In preparing this book I would like to acknowledge the help of others: Mike Dennison, G3XDV, Steve Telenius-Lowe, G4JVG, and Mike Prince, G7EUL.

Ian Poole
June 2001

CHAPTER 1

An introduction to the HF bands

In this chapter:

- Aspects of the hobby
- Amateur radio bands

THE short-wave (HF) bands are one of the main areas of activity within amateur radio today. On these bands it is possible to hear and contact stations from all over the world, and even a newcomer can make contacts with stations that are thousands of miles away.

Operating skills and know-how, combined with the right equipment, are all important to make the most of operating on these fascinating bands. While a newcomer will be able to make many interesting contacts, experienced operators tend to be far more successful, often making more exciting contacts with stations on unusual islands or on the other side of the globe. Much of this can be attributed to experience. This may be in assembling, maintaining and improving the station but it could be from knowing when to listen on the bands or which frequencies to use at a particular time. It might also be from knowing where to find the latest information about what is happening on the bands. Yet again it may be from knowing how to seek the stations out and how to make contacts when many others are calling, which are also very important. These and many other aspects are all very valuable, and are keys to gaining more enjoyment from operating.

Aspects of the hobby

There are many different ways in which people enjoy the HF bands. One of the major interests on them is *DXing* which is searching out stations from distant or interesting countries. It is quite possible to make contact with stations in virtually every country in the world and top DXers have scores in excess of 300. As a further challenge these countries can be contacted on several bands. People also enjoy making contact with stations on different islands.

DXers soon become highly skilled operators but they also have to ensure their stations are operating at maximum efficiency. This means that many people spend quite a bit of time improving their equipment, both inside and outside the shack.

At certain times of the year the amateur radio bands become alive with an enormous amount of activity as the result of a contest. These events take

HF AMATEUR RADIO

Station of a top UK DXer

a variety of forms but usually involve making as many contacts as possible, often with stations in a particular continent or country. They are great fun to join, and they also give the opportunity of making contact with stations from new or rare countries.

Some countries have little or no amateur radio activity and expeditions are often set up to activate them from an amateur radio viewpoint. Not only is this great fun for those operating the station, but it also gives the chance to visit some far-away and interesting countries. In many cases these *DXpeditions* are organised to coincide with one of the major contests to ensure they make the maximum number of contacts.

Once contacts have been made, stations often exchange *QSL cards* to confirm the contact. Collecting these from stations around the world can be a fascinating pastime of its own. In addition to this, some people like focussing some of their operating towards gaining operating awards. Once these have been obtained there can be a great sense of achievement.

Many operators enjoy talking to old friends around the world and the HF bands can provide the medium for keeping up with them. In fact some radio amateurs who have made friends with people overseas have regular contacts with them, possibly once a week, and chat to them about everything from how the family is keeping to what is happening in the world of amateur radio.

There is plenty of scope for experimentation on the HF bands. Antennas are a particularly popular area because even small improvements can greatly enhance the performance of the whole station. This means that time and

CHAPTER 1: AN INTRODUCTION TO THE HF BANDS

The D68C expedition to the Comoros Islands when over 160,000 contacts were made. Many of these contacts were made with people who had average or modest stations

energy spent on improving the antenna is always a wise investment and can result in the operator being able to contact an elusive or rare station. However, there are many other areas for experiments. Many people like to

A selection of QSL cards and awards

build some of their own equipment. Although most could not hope to produce equipment of the complexity and standard of commercially made receivers and transceivers manufactured by the large companies, there is still plenty of scope for the home constructor. There are lots of useful ancillary items of equipment that can be built relatively easily. Also there is a growing number of QRP (low-power) operators who use simple transmitters and receivers, many of which they build themselves. Not only is there a great challenge in building the equipment, but it develops operating skills because of the low powers that are used.

There is a variety of different types of modes of transmission that can be used. Single sideband is almost exclusively employed for speech transmissions. However, Morse (CW) is extensively utilised. To some it may seem surprising that it still has technical advantages over other transmission modes and can enable contacts to be made when using other modes would not succeed. Not only this, but many people enjoy using it! There are also data modes like radio teletype and the more up-to-date versions like AMTOR and PSK31. Slow-scan television is also used and provides a considerable amount of interest for some.

Another area in which the HF bands can be enjoyed is by helping push back the frontiers of technology. In the early days of amateur radio it was radio amateurs that discovered the value of the short-wave bands, and even today there is much work being undertaken to enable us to have a better understanding of the way in which signals propagate. There are other areas in which experimentation of this nature can be undertaken. For example, new forms of data transmission are being devised to enable data to be transmitted more accurately through interference or to give more facilities. These are just two of the ways in which people can turn their talents and amateur radio to help the development of technology and understanding.

CHAPTER 1: AN INTRODUCTION TO THE HF BANDS

Radio amateurs provided communications following the Lockerbie plane crash

Amateur radio can be used to help the community, especially after disasters have struck. Often it has been the only link for small islands hit by storms when all other forms of communications have been rendered useless. Similarly amateur radio has provided vital emergency communications after earthquakes and floods. The resilience and ingenuity of amateur operators under these circumstances has saved many lives. Whilst the UK is rarely hit by hurricanes, here too amateur radio has been used on many occasions to help those in distress.

The radio frequency spectrum extends over a vast range. It is split up into various areas to enable sections to be referenced more easily. Strictly speaking, the HF (high-frequency) portion covers 3.0 to 30MHz and includes those frequencies that are referred to as the *short-wave bands*. However, the short-wave bands may be considered as any bands having a frequency between the medium-wave broadcast band and 30MHz.

Within these limits there are a number of bands allocated to amateur radio. 'Top Band', covering 1.81 to 2.0MHz in the UK, is normally included and, although not strictly part of the HF section of the spectrum, it is considered as a short-wave band

Table 1.1. UK HF amateur radio bands

Frequency (MHz)	Wavelength (m)
1.810–2.000	160 (Top Band)
3.500–3.800	80
[3.500–4.000	75 – in the USA]
7.000–7.100	40
[7.000–7.300	40 – in the USA]
10.100–10.150	30
14.000–14.350	20
18.068–18.168	17
21.000–21.450	15
24.890–24.990	12
28.000–29.700	10

HF AMATEUR RADIO

Amateur radio proved the only communication when a hurricane hit Samoa

and will be included here. At the other end of the HF part of the spectrum is the 10m band. This extends to a frequency of 29.7MHz and is right at the opposite end of the short-wave spectrum.

With bands as diverse in frequency as these it is hardly surprising that there are huge differences in their characteristics. Some support long-distance propagation only during the day while others are best at night. There are many other differences, ranging from the distances that can be achieved to the size of the antennas that are used. This all adds to the variety that can be experienced in this part of the spectrum.

CHAPTER 2

Radio wave propagation

In this chapter:

- Ground waves and sky waves
- The atmosphere
- The ionosphere
- Layers in the ionosphere
- Variations in the ionosphere
- Definitions and terms
- Multiple hops
- Grey-line propagation
- Predicting conditions

ONE of the key skills available to any HF DXer is a knowledge of radio signal propagation. Knowing when to listen, which frequencies to use, where the signals may be coming from and many other points all give the experienced DXer a vital edge over other users. In fact, a good knowledge of propagation is an essential weapon in any DXer's armoury. Apart from this it is a particularly fascinating topic, and many people like to study it because it is so interesting, but to those who enjoy operating it is a key element in ensuring they can make contacts with the stations and countries they want.

There are many ways in which signals propagate. The most obvious is the way they would travel in free space where the signal spreads out in all directions. This can be likened to the ripples on a pond after a stone is dropped into the water. However, when signals are transmitted from a station on the Earth the signals are affected by the close proximity of the ground as well as other elements, including the ionosphere. In fact most signals on the MF and HF bands are heard via what is termed the *ground wave*, or after they have been refracted back to ground from the ionosphere (*sky wave*). As a result signals are able to travel further than the line-of-sight distance, and often as far as the other side of the globe.

Ground waves and sky waves

Ground waves occur as the signal spreads out from the transmitter. Instead of travelling in a direct straight line and not being heard beyond the horizon,

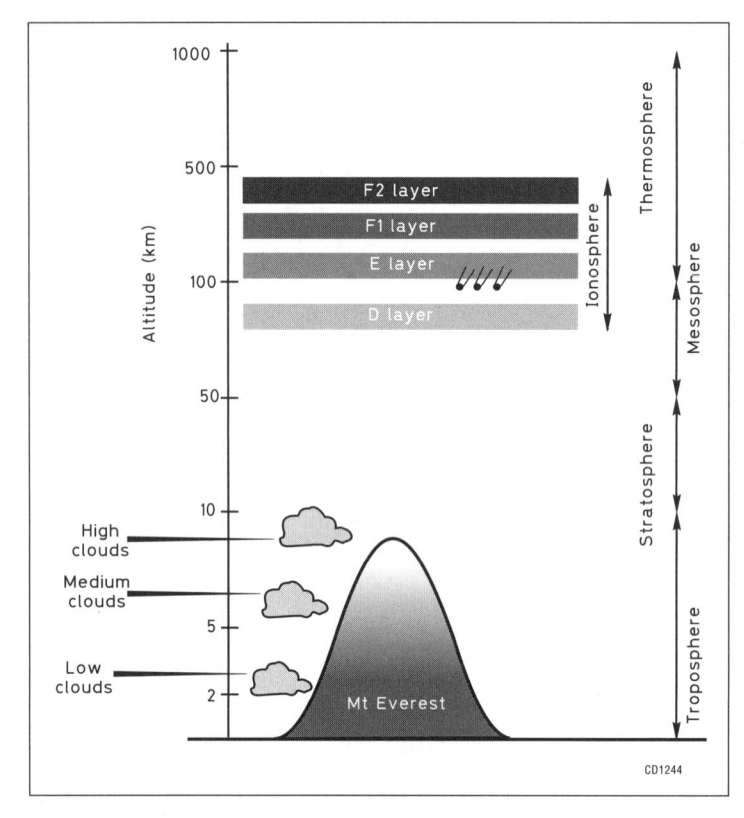

Fig 2.1. The composition of the atmosphere

the signal tends to follow the curvature of the Earth. The reason for this is that currents are induced in the surface of the Earth, slowing the wavefront close to it. This has the effect of tilting the wavefront downwards so that it follows the Earth's curvature and can be heard beyond the horizon.

It is found that the ground-wave signal becomes attenuated more at higher frequencies and the coverage is reduced. As a result it is not generally used for signals much above 2 or 3MHz.

A coverage area extending to 150km and more may be expected for a high-power broadcast station operating on the medium waveband, whereas a short-wave broadcast station will have a very small coverage area. These and other short-wave stations (including radio amateurs) rely on the *sky-wave* signals that travel away from the Earth and towards the ionosphere to give ranges of many thousands of miles.

The atmosphere

Before taking a look at how signals are reflected by the ionosphere it is very useful to find out a little more about the areas where the reflections take place and how these areas are formed. The atmosphere can be split up into a variety of different layers according to their properties. The most commonly used names arise from their meteorological properties.

The area closest to the Earth is known as the *troposphere*. This extends to an altitude of about 10km and it has an effect on propagation mainly in the VHF and UHF portions of the spectrum. It does not noticeably affect the short-wave bands. Above the troposphere is an area known as the *stratosphere*. This is found at altitudes between about 10 and 50km and contains the famous ozone layer at an altitude of about 20km. Next is the *mesosphere*, extending to an altitude of about 80km, and then on top of this is the *thermosphere* where temperatures can soar to 1200°C.

For short-wave communications an area known as the *ionosphere* is all-important. This crosses several of the meteorological boundaries and extends from an altitude of about 50km right up to about 600km.

The ionosphere

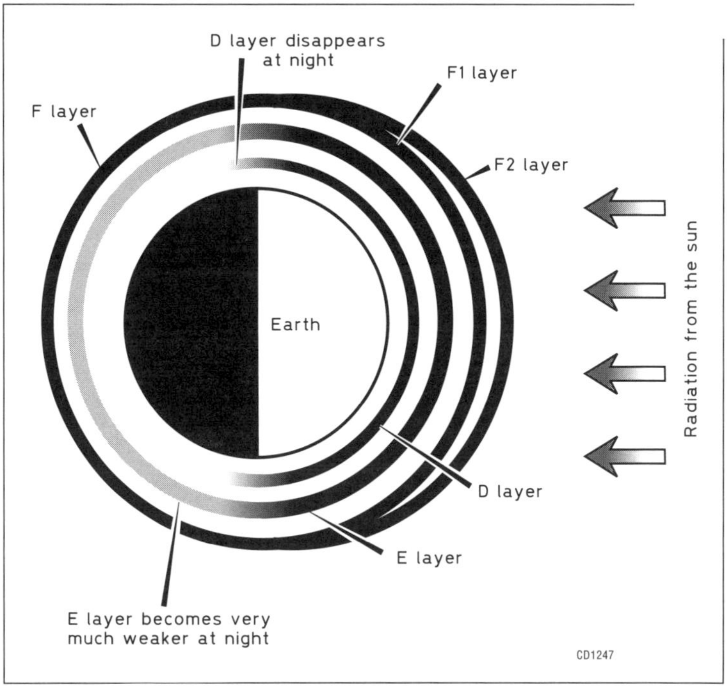

The ionosphere has been given this name because it is an area where chemical ions exist. In most areas of the atmosphere the gas molecules are in a combined state and are electrically neutral. However, in the ionosphere the gas molecules become ionised, forming a positive ion (a molecule that has lost an electron) and a free electron – it is actually these free electrons and not the positive ions that affect radio waves.

Ionisation occurs as a result of intense solar radiation at these altitudes splitting the gas molecules. Of all the forms of radiation from the Sun it is mainly the ultraviolet light that causes this effect. It can first be detected at an altitude of around 30km, but the electron density is not high enough to affect radio signals until an altitude of around 50km is reached.

Fig 2.2. Variations in the ionosphere over the period of a day

The ionosphere is often thought of as a number of distinct layers. While this is convenient for many explanations it is not strictly accurate because the whole of the ionosphere contains ionised molecules and free electrons. Instead the different layers can be visualised as peaks in the level of ionisation. There are three such peaks: the lowest is known as the *D layer*, above this is the *E layer* and still higher is the *F layer*. (Strictly speaking, there is also a *C layer* but the degree of ionisation there is so low that it has no effect on radio waves.)

The D layer

The D layer is found at altitudes between 50 and 80km. It is only present during the day when the Sun's radiation is present. The reason for this is that free electrons and positive ions recombine to form neutral molecules. At this altitude the air density is relatively high and recombination occurs quite quickly – as a result the Sun's radiation is needed to retain the level of ionisation. When it is removed the level of ionisation quickly falls and the layer effectively disappears at night.

The D layer acts as an attenuator to radio signals, and this is particularly noticeable at low frequencies. This can be seen by the fact that signals in the medium-wave broadcast band are prevented from reaching the higher layers during the day and they are not heard beyond the range of the ground wave. It is found that the level of attenuation varies according to an inverse square law. In other words, if the frequency doubles, the level of attenuation falls by a factor of four.

The signals are attenuated because the electrons in the layer are vibrating in synchronism with the frequency of the signal. The air density is still relatively high at this altitude and the vibrating electrons collide with other air molecules. A small amount of energy is lost at each collision and the signal is reduced.

The actual level of attenuation is related to the number of collisions that take place. This is obviously dependent upon the level of ionisation but it is also dependent upon the frequency of the radio signal. As the frequency increases, so the wavelength of the vibrations decreases and the number of collisions falls.

As a result low-frequency signals are attenuated more than higher-frequency ones, although it should be remembered that high-frequency signals still suffer some attenuation.

The E layer

The E layer appears above the D layer and can be found at altitudes between about 100 and 125km. Here, too, the ions combine relatively quickly and after dark the level of ionisation falls quite rapidly. Although a small amount of residual ionisation remains after dark it virtually disappears.

When signals enter the E layer they cause the electrons to vibrate. However, the air density is much less than at the altitude of the D layer and there are far fewer collisions. As a result much less energy is lost and the layer affects radio waves in a different way. Unlike the D layer where the radio signal causes the electrons to vibrate and collide with other gas molecules, the lower gas density means that when the electrons vibrate, the signal is actually re-radiated. Because the signal is travelling into an area where the level of ionisation (and hence the number of free electrons) is increasing, it is found that this has the effect of bending or *refracting* it away from the area of higher electron density. This refraction is often sufficient at frequencies in the HF bands to bend the signal back to Earth, as if it had been reflected.

These 'reflections' are affected by the frequency of the signal and the angle at which the signal enters the layer. It is found that as the frequency increases, so the amount of refraction decreases and a point is reached where the signals pass straight through.

The F layer

The highest and most important of the ionospheric layers for long-distance communication is the F layer. During the day it often splits into two sub-layers that are called the *F1* and *F2 layers*. Then at night they merge back into a single layer.

The height of the layers varies considerably according to the time of day, the season and the state of the Sun. However, as a rough guide, in summer the F1 layer may be at an altitude centred around 300km and the F2 layer centred around 400km. In winter these figures may fall to around 200 and

300km. At night the combined F layer is generally between about 250 and 300km.

The level of ionisation of the F layer falls just like the other layers but, because the level of ionisation is much greater, there is still sufficient ionisation present to affect signals at night. Like the E layer, the F layer acts as a reflector of signals rather than an attenuator. Also, it is the highest of the layers and so the distances that can be achieved using a single reflection from it are the greatest.

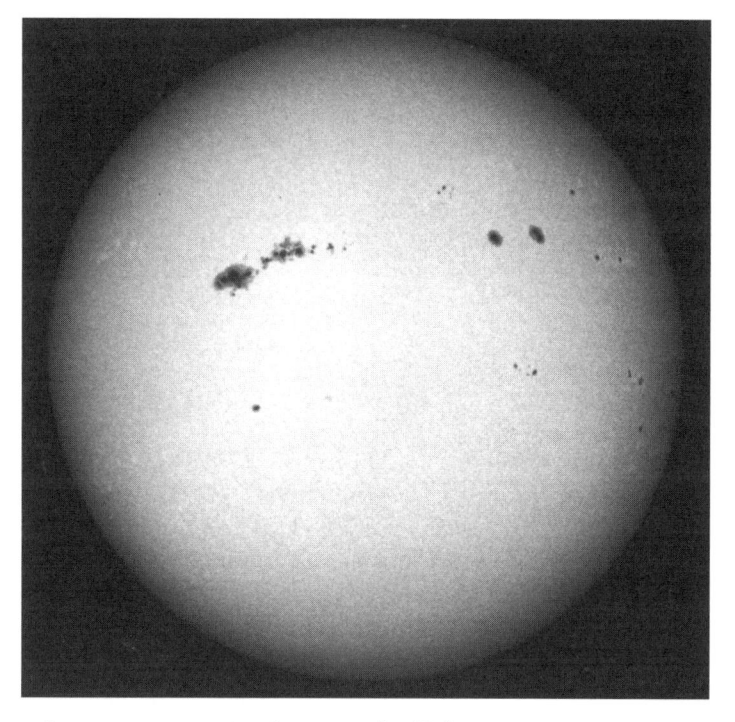

The Sun, showing several sunspots

Varying the frequency

During the day it is found that signals on the medium-wave band are only able to use ground-wave propagation because the D layer absorbs any sky-wave signals. At night, when it disappears, signals can be reflected back to Earth.

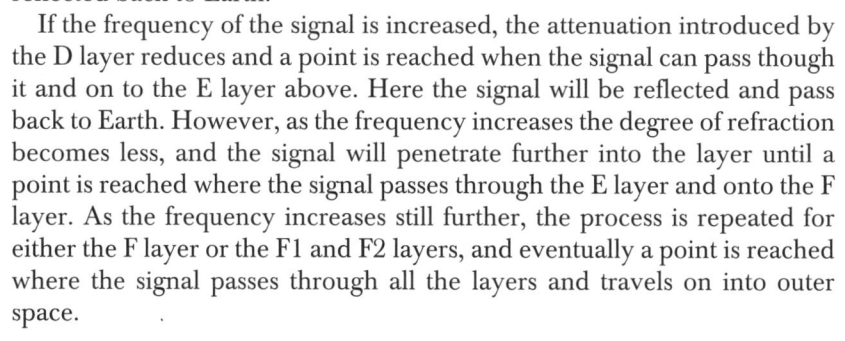

If the frequency of the signal is increased, the attenuation introduced by the D layer reduces and a point is reached when the signal can pass though it and on to the E layer above. Here the signal will be reflected and pass back to Earth. However, as the frequency increases the degree of refraction becomes less, and the signal will penetrate further into the layer until a point is reached where the signal passes through the E layer and onto the F layer. As the frequency increases still further, the process is repeated for either the F layer or the F1 and F2 layers, and eventually a point is reached where the signal passes through all the layers and travels on into outer space.

Variations of the ionosphere

The state of the ionosphere is constantly changing and this results in changing conditions on the HF bands. This is dependent upon the amount of radiation received from the Sun, and hence the time of day has a major effect. So does the season – in the same way that more heat is received from the Sun in summer, more radiation hits the ionosphere. Furthermore, the actual state of the Sun has a major effect, in particular the number of *sunspots* on its surface. These spots appear as areas that are comparatively dark when compared to the rest of the surface. However, they affect the ionosphere because the areas around the spots emit greater amounts of ultraviolet light and that is the main cause of ionisation.

Fig 2.3. The angle of radiation, showing how low-angle signals travel further

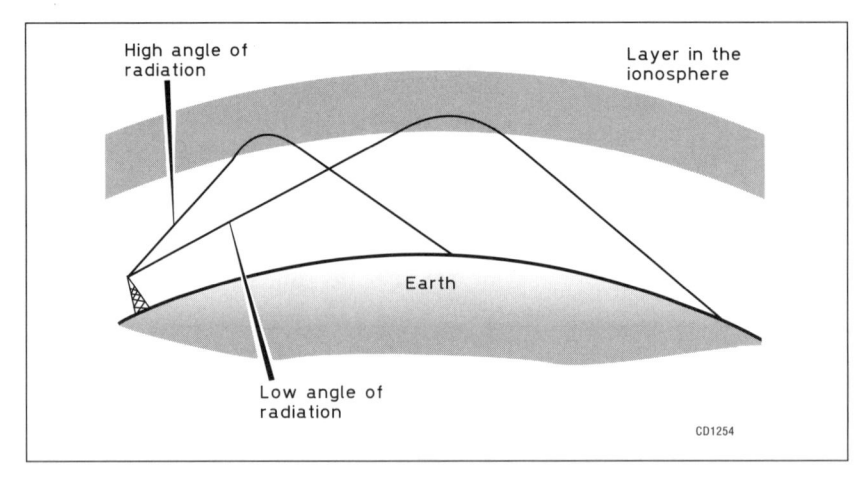

Astronomers have observed sunspots and it has been found that the number of them varies cyclically with a period of around 11 years. As a result ionospheric conditions and hence radio propagation also vary in line with this cycle. Broadly speaking, at the low point of the cycle the HF bands above about 20MHz may not support ionospheric propagation, while at or near the peak of the cycle frequencies of 50MHz and higher may be affected.

Definitions and terms

When talking about the ionosphere there are a number of terms that are commonly used. The first is the signal's *angle of radiation*. This is effectively the angle between the main beam of the signal and the ground. Very low angles of radiation travel almost parallel to the Earth at first. High angles of radiation travel upwards towards the ionosphere as shown in Fig 2.3 and hit it with a high angle of incidence. They will need to receive a higher degree of refraction to be returned to the Earth, and are therefore more likely to pass through a layer than ones with a lower angle of incidence. Additionally those signals that have a low angle of radiation are able to travel further as a result of the geometry.

Another term is the *skip distance*. This is the distance the signal travels along the surface of the Earth as shown in Fig 2.4. The maximum distance that can be reached using a single refraction is about 2500km for the E layer and 5000km for the F layer. It is important to note that this can only be achieved if a very low angle of radiation is used. Even with an angle of radiation of 20° the maximum distances fall to 400 and 1000km respectively.

There is also a region in which no signal may be heard. This occurs in the region after the ground wave has been attenuated to the extent it cannot be heard, and before the first sky-wave signals are returned to Earth. This is known as the *skip zone* or *dead zone*.

Naturally signals can be heard over greater distances. This is achieved by a sky wave being refracted back to Earth and then being reflected back up

CHAPTER 2: RADIO WAVE PROPAGATION

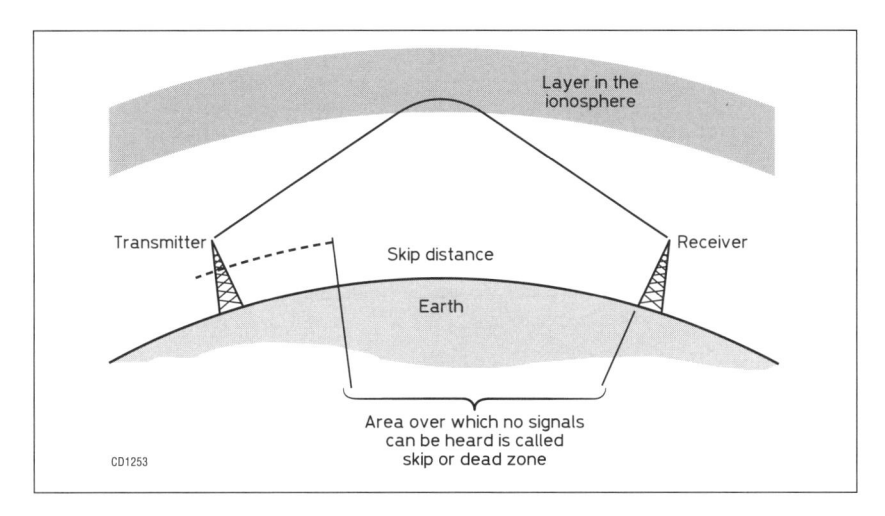

Fig 2.4. Skip distance and dead zone

to the ionosphere a second time (see Fig 2.5). In this way the path includes two hops. However, as the signal has to pass through the D layer more times, the signal strengths are usually lower.

The *critical frequency* is another term that is often encountered. One of the ways of measuring the state of the ionosphere is to send a series of pulses directly upwards. This is known as *ionospheric sounding*. If the frequency of the transmitted pulses gradually increases, it is found that at first the pulses are returned to the Earth. As the frequency is increased, the signal penetrates further into the layer and eventually a point comes where it passes through it. The point at which the signal just passes through a layer and on to the next is called the *critical frequency* for that layer. Also, by measuring the time taken for a pulse to be returned it is possible to measure the effective height of a given layer.

Another term, the *lowest usable frequency* (LUF), is very important. It is found that as the frequency of a transmission is reduced, the losses increase, and a point is reached where the signal cannot be copied. The LUF is defined as the frequency where the signal equals the minimum strength for satisfactory reception.

There is also a term called the *maximum usable frequency* (MUF), which is

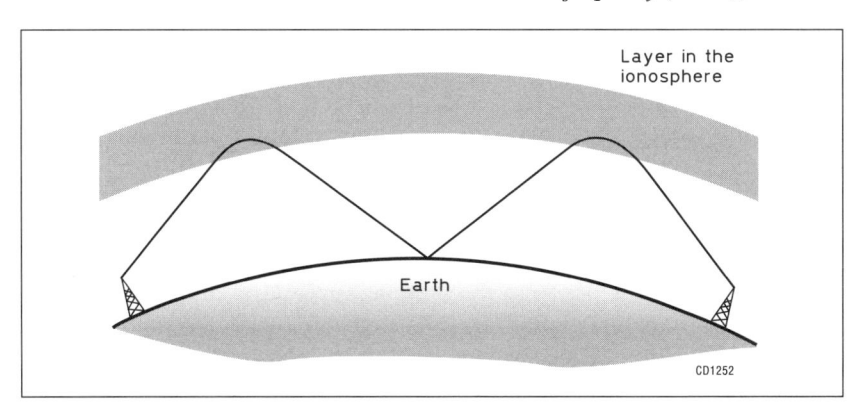

Fig 2.5. Multiple reflections from the ionosphere

13

more relevant to the other end of the spectrum. As the frequency of a signal is increased it penetrates further into the layers and eventually passes right through. A point is reached where communications just start to fail. This is the maximum usable frequency. Generally the MUF is between three and five times the critical frequency.

Ionospheric storms

On occasions the normal propagation on the HF bands can be disrupted. Sometimes this may be for a few hours, but other times it may last for a few days. There are a number of reasons for this, but one of the main causes is a *flare* occurring on the Sun. When this happens, huge amounts of material erupt from under the surface of the Sun and are thrown out into space. Some of the particles emitted travel at a colossal velocity – about a tenth the speed of light, while other particles are somewhat slower, travelling at only about 1000km per second. Along with the eruption, vast amounts of radiation are also emitted. When seen using the right equipment the flare looks like an enormous flame leaving the surface of the Sun, and this can last for about an hour, after which the surface settles back to its former state.

There are a number of effects of a flare on radio communications. The first one is seen when the radiation arrives. The level of ionisation in the D layer rises very sharply, often absorbing signals throughout the HF spectrum, although this only occurs on the sunlight side of the Earth. This effect is known as a *sudden ionospheric disturbance* (SID) and may last for a few hours. Dependent upon its severity it may only affect the lower frequencies.

It takes a few hours for the high-energy particles to arrive. When they do, they are deflected by the Earth's magnetic field so that they move towards the poles where they cause a very large increase in the level of D-layer absorption. This can last for up to three or four days, and during this time such *polar cap absorption* can prevent HF communications across the poles.

Between 20 and 40 hours later the lower-energy particles arrive. These cause a change in the Earth's magnetic activity and this is recorded by changes in the A and K indices which act as an indicator of the stability of the HF band conditions. A *geomagnetic storm* may give rise to an *ionospheric storm*. If this occurs the chemistry of the ionosphere is affected, depressing the levels of ionisation by a significant degree and reducing the frequencies that can be reflected by the E and F layers. At the same time there is an increase in the level of absorption. The total effect is that the maximum usable frequency falls and the lowest usable frequency rises, reducing the band of frequencies that can be used. In some instances the MUF and LUF may meet and there will be a total radio blackout. In this case the only stations that can be heard are via ground wave. The effects may last for up to a week for a bad storm, with the bands slowly returning to normal.

Grey line

One form of propagation that can yield exceedingly good results is known as *grey-line propagation*. It occurs when signals travel in the twilight (grey-line) zone. It is found that propagation there is very efficient because the losses are

CHAPTER 2: RADIO WAVE PROPAGATION

very much lower. The reason is the level of ionisation in the D layer is very much reduced while the ionisation in the F layer (by which most reflection occurs) has not decayed very much.

To utilise this mode of propagation both the transmitting and receiving stations must be on the grey line or twilight zone. The latter does not last for long, typically only remaining effective for about 30 minutes either side of sunrise or sunset.

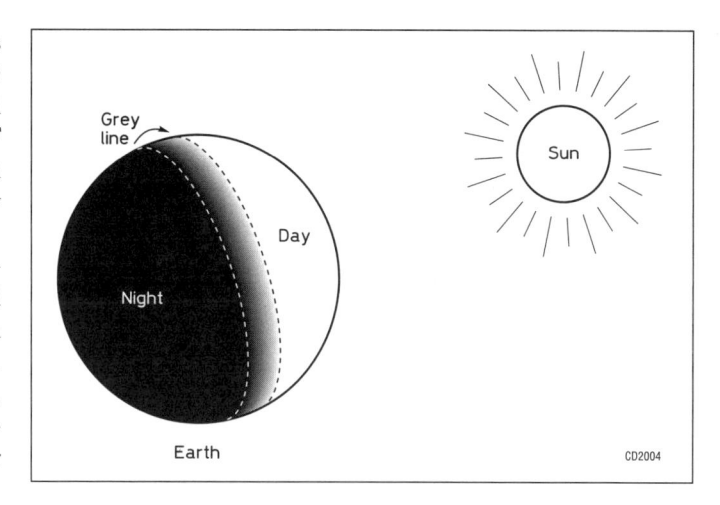

Fig 2.7. The grey-line (twilight) zone. This is an area about 30 minutes either side of sunrise or sunset along which radio signals propagate with much-reduced losses

The mode can be at its most dramatic on the LF bands and signals from the other side of the globe can be heard rapidly rising out of the noise to become as strong as many locals. Then later, when the grey-line enhancement has passed, they fall away again just as quickly.

During the course of a year the grey line changes as shown in Fig 2.8 overleaf. This means that the areas of the globe from which stations can be heard using this form of propagation vary, as do the times of day when it occurs.

Predicting conditions

Knowing how the different bands perform and how to predict and use the different conditions that occur is one of the skills that is essential for anyone interested in the HF bands. In many respects, predicting what the propagation conditions are going to be like is rather akin to weather forecasting.

Propagation is affected by a number of factors – frequency, time of day, the season, position in the sunspot cycle, and general state of the Sun all affect the way in which the signals can propagate.

The time of day has a significant effect. At night, when no radiation is received from the Sun, the level of ionisation reduces. On low frequencies this means that signals can reach the reflecting layers and they can be heard over much greater distances. However, at higher frequencies the reduction in the level of ionisation means that signals at frequencies that would have been reflected may pass right through the ionosphere. In other words the maximum usable frequency is reduced.

Similar effects occur at the different seasons. In winter the level of radiation received in a particular hemisphere is reduced in the same way that the Sun's rays do not provide as much warmth. This means that the levels of ionisation are reduced. The low-frequency bands open earlier and perform better, the maximum usable frequencies are less, and the high-frequency bands close earlier in the evening.

The sunspot cycle also has an effect. At the low point of the cycle, frequencies below 20MHz may not support propagation via the ionosphere

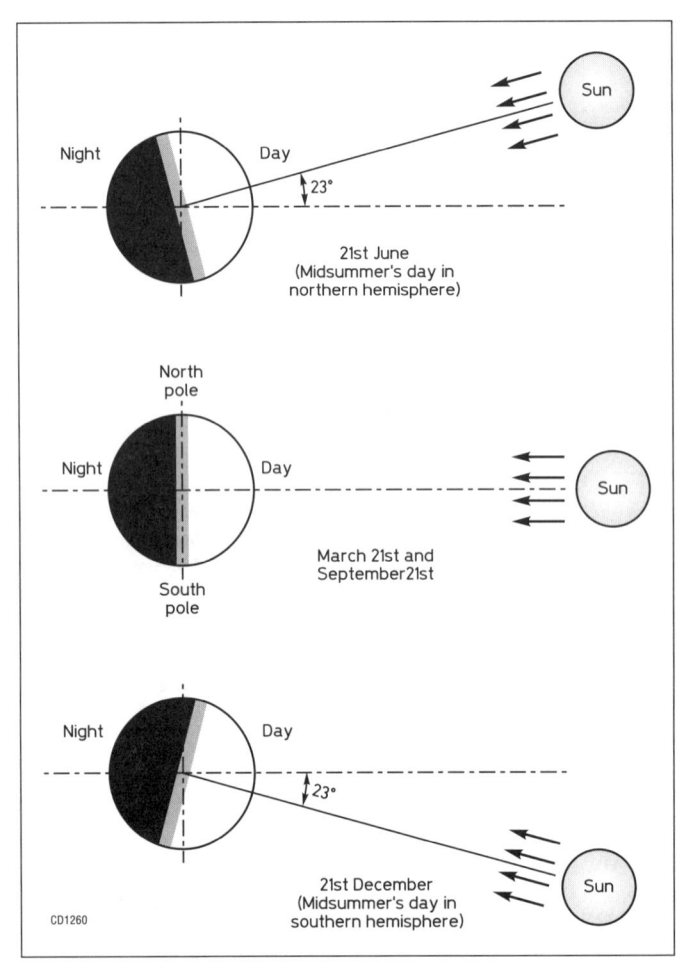

Fig 2.8. Grey line changes with the season

whereas at the peak of the cycle frequencies in excess of 50MHz may be reflected.

To be able to estimate what the conditions may be like, figures are available that indicate the effects of the Sun. The first is the *solar flux* – this is an indication of the amount of radiation being received from the Sun. It is found that the latter not only emits vast quantities of heat and ultraviolet radiation but also radio energy. The solar flux is measured at a frequency of 2800MHz (10.8cm) and is given in *solar flux units* (SFU). The actual measurement is made at the Penticton Radio Observatory in British Columbia, Canada, at 1700 UTC each day. The value can vary from around 60 up to 300 and beyond. The higher the value, the better the likelihood of good conditions.

Other factors are also involved, including the magnetic activity. An index known as the *A index* reflects the severity of the magnetic flux occurring at local magnetic monitoring points around the globe. During magnetic storms the index may reach levels of 100 and, during severe storms, as high as 200 and more. As the A index varies from one point on the globe to another an index known as the *Ap index* is usually quoted as it is the 'planetary' index.

The *K index* is more often used and ranges from 0 to 9. The A index has a 24-hour format while the K index is updated every three hours. It is related to the A index as shown in Table 2.1, and Table 2.2 shows the levels of the associated storms.

As a rough guide the solar flux should be at around 150 to produce good conditions on the HF bands. This level should be maintained for a few days, as it take several days of high solar flux levels to improve the conditions. In conjunction with this the geomagnetic activity should be low with K index levels of between 0 and 2.

Apart from gaining a good idea of what conditions will be like from an estimate using these figures, it is also possible

Table 2.1. Conversion between A and K indices

K index	A index
0	0
1	3
2	7
3	15
4	27
5	48
6	80
7	140
8	240
9	400

to utilise propagation software. Some of these programs are very sophisticated, being originally developed by broadcast or military organisations. They include ASAPS, VOACAP, and many others. Some of them are now available for comparatively little cost and can be used on a PC in the shack. It is not intended to review them here as they are constantly changing and being updated. Some may also be given away as 'free' software with the radio magazines from time to time, giving an ideal chance to evaluate the most convenient one for a given situation.

Table 2.2. Geomagnetic activity indices

Conditions	A Index	Potential of a storm
Quiet	0–7	Low
Unsettled	8–15	Low
Active	16–29	Moderate
Minor storm	30–49	Moderate
Major storm	50–99	High
Severe storm	>100	Very high

Further reading
Your Guide to Propagation, Ian Poole, RSGB, 1998.

CHAPTER 3

Types of transmission

In this chapter:

- Morse
- Amplitude modulation
- Single sideband
- Frequency modulation
- Radio teletype
- AMTOR
- PSK31
- Slow-scan television

A GREAT VARIETY of types of transmission can be heard on the HF bands, both inside and outside the amateur bands. A few of these transmissions may be intelligible, carrying speech or music, whereas others may be Morse signals. Some signals may appear to be using another form of keying and others might even sound more like computer data.

The reason for using all these types of transmission is that they enable communication to be made in many different ways. Each type has its own advantages and may be used under different circumstances. Within amateur radio, several types of transmission are allowed: Morse, amplitude modulation, radio teletype, data, slow scan television, to name a few. This gives a tremendous amount of variety and flexibility in amateur communications.

A radio signal consists of two main components: the carrier and the modulation. The *carrier* is a steady-state signal which is then modified (*modulated*) in the transmitter so that it can carry information. When the resulting signal is received, it has to be *demodulated* so that the information (*modulation*) is removed from the carrier. If this information consists of sound waves it may then be passed through the audio amplifier and on to a loudspeaker or headphones to be heard in the normal way. Alternatively this demodulated signal may be passed into a computer for processing if it consists of data.

Morse

One of the oldest, simplest (but still one of the most effective) forms of modulation is Morse code (CW). It has been used for transmissions since

HF AMATEUR RADIO

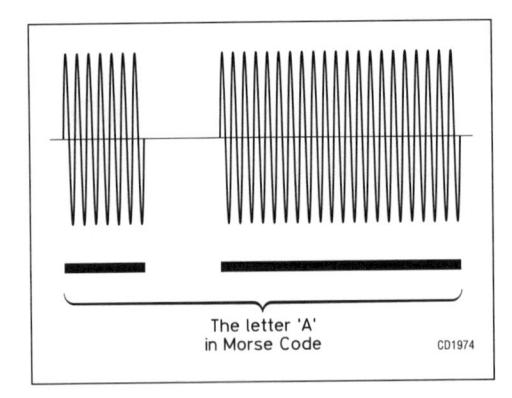

Fig 3.1. A Morse signal as displayed by an oscilloscope

The letter 'A' in Morse Code

the very earliest days of radio, and was originally used for telegraph communications from the middle of the nineteenth century onwards. Despite its age it still has a very important place in today's high-technology radio scene.

Possibly one of the most obvious advantages is that Morse can be transmitted on very simple equipment. All that is needed is a circuit that generates a radio-frequency signal and a means of turning it on and off. In fact a simple transmitter can be made from as few as two or three transistors and a handful of other components.

Morse has a number of technical advantages as well. The relatively low modulation rate means that the transmission only occupies a small bandwidth, and as a result the filter bandwidths required to receive the signal can be made very narrow: 500, 250Hz or less. These bandwidths are much narrower than that required for single sideband speech, which is around 2.5kHz, and using them reduces both the level of interference from other signals and the level of background noise.

A further advantage of Morse is that the human brain can decipher Morse signals at very low levels because it simply consists of a signal that is turned on and off. They can even be copied when they are below the noise level, whereas a single sideband signal must be above the noise level to be copied. Both these factors give Morse an advantage of more than 10dB (10 times) over other modes. As a result of this it can be used to make contact when other modes would not succeed, and many stations, especially those with average antennas and transmitter power levels, use Morse because it enables them to make many more contacts. This is particularly true for those who like to use low power.

To resolve a Morse signal the receiver needs to have a beat frequency oscillator (BFO) as described in Chapter 4. This converts the on-off keying of the carrier into the characteristic note associated with Morse code. All communications receivers and most 'world band' receivers have a beat frequency oscillator included, activated from part of the mode switch or from a separate BFO on/off switch. In the absence of a Morse or CW position on a mode switch, the SSB position can be selected since a BFO is also required for resolving this mode.

The main disadvantage of Morse is the time required to learn the code. While this may appear to be tiresome, people with an average station find that Morse provides an excellent way of making long-distance contacts. It is for this reason, together with the simplicity of the equipment, that many low-power (QRP) operators use it almost exclusively. Even those with very good stations use it widely to ensure they maximise their opportunities of contacting all the stations they need. As a result Morse is still very popular on the amateur bands and this can be verified very quickly by listening in the Morse sections of them. In fact it is often possible to hear Morse stations on some bands when no other signals are present.

CHAPTER 3: TYPES OF TRANSMISSION

An old hand key and a paddle designed for use with a modern electronic keyer

Amplitude modulation

Even though Morse possesses many advantages it cannot convey music or the spoken word. To achieve this the carrier must be modulated to follow the variations in sound intensity. The simplest method of achieving this is

HF AMATEUR RADIO

to vary the amplitude of the signal in line with the variations in the audio signal. This form of modulation is known as *amplitude modulation* (AM) and it is widely used for broadcasting on the long-, medium- and short-wave bands. It is also used on the aircraft band above 108MHz but it is not used to any degree by radio amateurs.

The chief advantage of amplitude modulation is its simplicity. Demodulation can be undertaken using a diode and a couple of other components. This means that AM receivers can be made cheaply and easily, a factor demonstrated by the availability and the low cost of many AM-only radios.

However, AM does not use the power or spectrum bandwidth it requires very efficiently. To see the reason for this it is necessary to look at the background. Fig 3.2 (a) shows an unmodulated carrier. The instantaneous value of the signal is seen to vary (Fig 3.2 (b)) when modulation is applied. The maximum level of modulation that can be achieved is when the envelope falls to zero and rises to twice the steady state value (Fig 3.2 (c)). When this occurs the signal is said to have *100% modulation*.

Looking at the frequency spectrum of the signal, it can be seen that if the carrier is modulated by a single tone, two other signals (*sidebands*) appear, one on each side of the main carrier (Fig 3.3). A 1kHz tone will produce sidebands spaced 1kHz from the carrier. It is found that under conditions of 100% modulation the maximum level of the sidebands rises to 50% of that of the main carrier, and this means that the power level of each sideband is only a quarter that of the carrier.

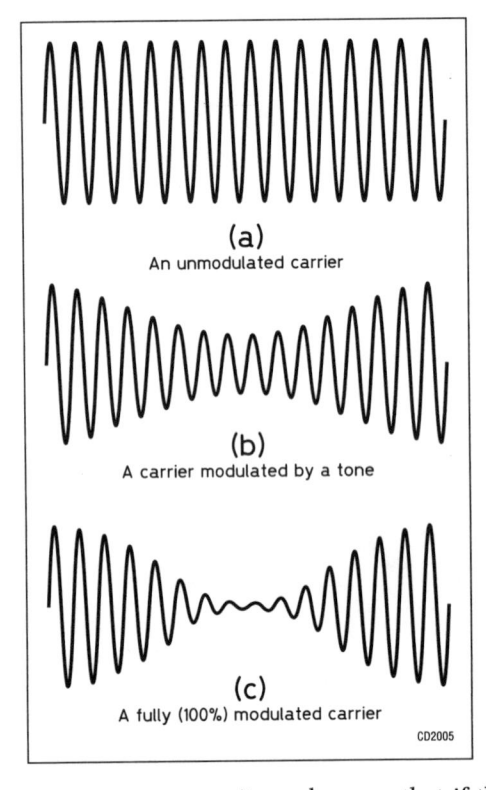

Fig 3.2. Amplitude modulation of a carrier by a sine-wave tone

When speech or music is applied instead of a steady tone, sidebands that stretch out either side of the carrier are seen (Fig 3.4). Again the maximum power of these is only a quarter of that of the main carrier. As the actual information for the audio is carried in the sidebands and the carrier only acts as a reference for demodulation, it can be seen that the system is not very efficient. Also, the bandwidth occupied is twice that of the original audio. These factors mean that it is not used for communications on the HF bands.

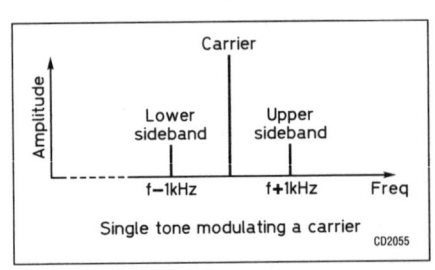

Fig 3.3. Spectrum of a carrier modulated by a 1kHz tone

Single sideband

For communications purposes another type of modulation derived from AM is used. Called *single sideband* (SSB), it sounds garbled when received by an ordinary receiver, but it makes far more efficient use of the available power and can be received at lower signal strengths than AM. In addition to this it occupies only half the bandwidth of an AM signal. In view of these advantages it is the mode that is used almost exclusively for speech communications on the HF bands, especially by radio amateurs.

The signal is derived from AM by removing the carrier and one of the sidebands (Fig 3.5). This is performed in the transmitter and can be achieved in a number of ways. The most obvious is to use filters but it is also possible to use phasing techniques. Normally the filter method produces the best results. Here a fixed-frequency carrier is applied to a double-balanced mixer along with the audio to produce a signal consisting of two sidebands and no carrier. A filter is then used to remove the unwanted sideband. Once this signal is generated, mixers are used to bring it to the required frequency.

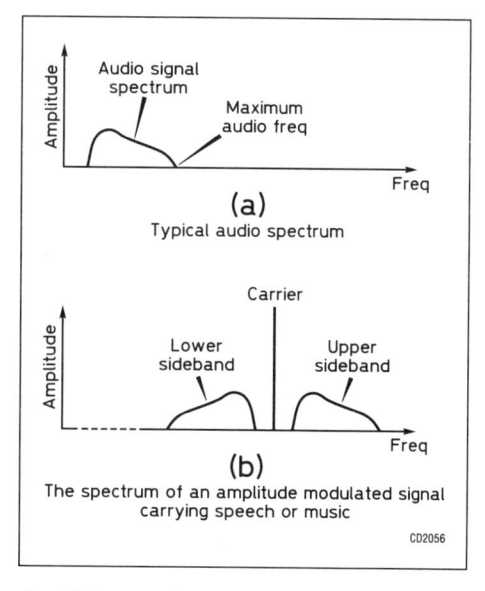

Fig 3.4. Spectrum of a carrier modulated by an audio signal

A BFO and a mixer are required in the receiver for SSB signals to be resolved. Sometimes the mixer may be called a *product detector* because the output is the product of the two input voltages at any instant. This term was used mainly with valve sets some years ago and is less commonly used these days, although it may still be encountered from time to time.

When the carrier is reinserted by the BFO it must be on approximately the correct frequency. Any error will manifest itself by raising or lowering the pitch of the reconstituted audio. For amateur communications a tolerance of around 100Hz can be accommodated, although any offset will give the reconstituted audio rather a peculiar 'tone'. For most receivers it is possible to achieve much closer tuning than this and the audio should sound reasonable.

When transmitting or receiving SSB it is necessary to know which sideband to use: either the upper sideband (USB) or the lower sideband (LSB). The convention for radio amateurs is that the LSB is used on frequencies below 10MHz and the USB above 10MHz.

Many stations use various forms of speech processing and find them particularly effective with SSB. They raise the average power level of the transmission and make the most effective use of the radiated power by ensuring that only the frequencies that contribute to the intelligibility are transmitted. Full details of these systems are found in Chapter 5 – 'Transmitters'.

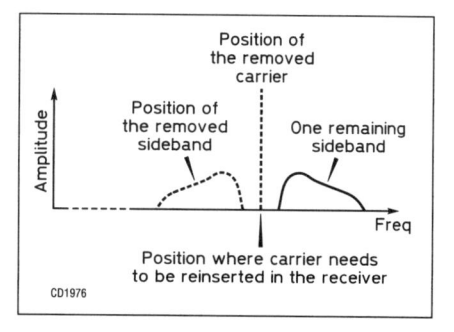

Fig 3.5. A single sideband signal consists simply of one sideband. The carrier and one sideband are removed

Frequency modulation

Although the most obvious way to modulate a signal is to vary its amplitude, there are other methods that can be used. One of these is to modulate its frequency. In this way the voltage of the modulating signal will determine the exact frequency of the carrier as shown in Fig 3.6. The rate of change of the carrier frequency is at the modulating frequency, and the amount of change is proportional to the amplitude of the modulation.

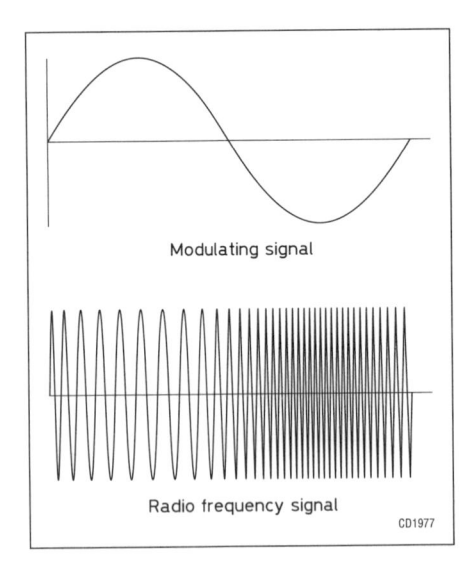

Demodulating FM requires a circuit that is frequency sensitive. In this the variations in frequency of the incoming signal are converted into variations in voltage at the output of the demodulator. There are a number of circuits that can be used to achieve this – many directly utilise a tuned circuit, but others are based around the use of a phase-locked loop. The phase-locked loop variants have the advantage that they normally do not use an inductor. As inductors can be difficult or expensive to produce, and phase-locked loops provide a very high level of performance, their use is often preferred.

The degree of frequency variation (*deviation*) of a frequency-modulated signal is obviously important. It should obviously not pass outside the bandwidth of the receiver. The transmissions used for broadcasting on the VHF bands use *wide-band frequency modulation* (WBFM) with a deviation of ±75kHz and a bandwidth of 200kHz. Transmissions used for communications purposes have much narrower deviation, typically ±3kHz, and are known as *narrow-band frequency modulation* (NBFM) signals.

One of the advantages of FM is that the modulation is carried as variations in frequency. A receiver can be made sensitive to the frequency variations and insensitive to any amplitude variations. Most noise appears as amplitude variations so the receiver can be made much less sensitive to this. The insensitivity to amplitude variations also means that variations in signal level, due to fading for example, will not be as noticeable as with an amplitude-modulated signal. Furthermore, if there is an interfering signal on the channel the receiver will tend to receive only the strongest one – a feature known as the *capture effect*.

Frequency modulation is mainly used for point-to-point communications on the VHF and UHF bands. It is also the preferred mode for mobile and local amateur communications at these frequencies. However, it is not widely used on the HF bands except at the top of the 10m band, above 29.5MHz, where it is used for local communications with the possibility of some DX contacts. There are even repeaters, especially in the USA, that help extend the coverage for those with poor antennas and locations. It is not unknown for people in their back yard, using a hand-held transceiver, to make worldwide contacts through these repeaters. Simplex operation is generally above 29.6MHz, and repeater outputs are on 29.62, 29.64, 29.66 and 29.68MHz with the input channels 100kHz lower.

Fig 3.6. Frequency modulation

Frequency shift keying

The revolution in digital information has touched most areas of technology and radio, and amateur radio is no exception. It is quite easy to transmit data over a radio link but appropriate forms of modulation need to be used. A system known as *frequency shift keying* (FSK) is used for many amateur data transmissions. Here the signal is continuously changed from one frequency to another, one frequency representing the digital one (*mark*) and the other the digital zero (*space*). By doing this in step with the digital stream of ones and zeroes it is possible to send data over the radio.

FSK is widely used on the HF bands. To generate the audio tone required from the receiver, a beat frequency oscillator must be used. Accordingly, to obtain the correct audio tones the receiver must be tuned to the correct frequency.

At frequencies in the VHF and UHF portion of the spectrum a slightly different approach is adopted. An audio tone is used to modulate the carrier and the audio is shifted between the two frequencies. Although the carrier can be amplitude modulated, frequency modulation is virtually standard. Using this *audio frequency shift keying* (AFSK) the tuning of the receiver becomes less critical.

When the data signal leaves the receiver it is generally in the form of an audio signal switching between two tones. It needs to be converted into the two digital signal levels, and this is achieved by a unit called a *modem* (modulator/demodulator). Audio tones fed into its receiver generate the digital levels required for a computer or other equipment to convert into legible text. Conversely it can convert the out-going digital signals into the audio tones required to modulate the transmitter. To be able to use these digital modes several stages of conversion are needed. The audio tones are applied to the modem but some form of decoding is then required to convert them to a form that can be displayed. Some processing is also required to cater for the protocols that are required with some of the more sophisticated modes used today, and finally a display and keyboard are required. Many stations use a piece of equipment known as a *terminal node controller* (TNC) to perform the modem function and to undertake the processing required for the protocols and transmit-receive switching. A computer or terminal is then required to display the data. However, the most popular method is to utilise a computer with suitable software and feed the tones into the audio card. In this way the computer is able to handle all the functions required without the need for any further equipment. As most amateur radio stations have access to a reasonable computer this is the most popular option.

The speed of the transmission is important. For the receiver to be able to decode the signal it must be expecting data at the same rate that it is arriving. Accordingly a number of standard speeds are used. These are normally given as a certain number of *baud* where one baud is equal to one bit per second.

Radio teletype

The first form of data transmission to gain widespread acceptance was known as *radio teletype* (RTTY). It was widely used commercially, and it soon became popular with radio amateurs when equipment became available on

Table 3.1. Murray or Baudot code

Lower case	Upper case	Code element 5 (MSB)	4	3	2	1	Decimal value
A	-	0	0	0	1	1	3
B	?	1	1	0	0	1	25
C	:	0	1	1	1	0	14
D	$ AB	0	1	0	0	1	9
E	3	0	0	0	0	1	1
F	! %	0	1	1	0	1	13
G	& @	1	1	0	1	0	26
H	£	1	0	1	0	0	20
I	8	0	0	1	1	0	6
J	' Bell	0	1	0	1	1	11
K	(	0	1	1	1	1	15
L	)	1	0	0	1	0	18
M	.	1	1	1	0	0	28
N	,	0	1	1	0	0	12
O	9	1	1	0	0	0	24
P	0	1	0	1	1	0	22
Q	1	1	0	1	1	1	23
R	4	0	1	0	1	0	10
S	Bell !	0	0	1	0	1	5
T	5	1	0	0	0	0	16
U	7	0	0	1	1	1	7
V	; =	1	1	1	1	0	30
W	2	1	0	0	1	1	19
X	1	1	1	1	0	1	29
Z	" +	1	0	0	0	1	17
Space		0	0	1	0	0	4
CR		0	1	0	0	0	8
LF		0	0	0	1	0	2
Figure shift		1	1	0	1	1	27
Letter shift		1	1	1	1	1	31
Blank		0	0	0	0	0	0

AB = Answer back or WHU (Who Are You?)
Upper-case characters may vary in some cases as indicated in the table. Also upper-case F, G, and H are not often used.

the second-hand market. Large mechanical teleprinters were used to print the data, although nowadays computers are used because they are far quieter and more effective.

Data is sent as a series of pulses, each character consisting of five bits of either a mark or a space. The actual code that is used is called the *Murray code* or *Baudot code* and this is shown in Table 3.1. This code is internationally recognised and to accommodate the various international requirements a number of national variations exist, although they are relatively minor.

The letter shift and figure shift codes are sent to change from upper case to lower case and vice versa. Once the code has been sent then the system will remain in that case until the next case-change code is sent. One of the drawbacks of RTTY is the limited set of characters that can be sent – only text and numbers, and very few other characters.

Data is sent relatively slowly because the mechanical teleprinters could not cope with data any faster. 45.5 baud is the standard amateur speed for HF, although other standards at 50, 56.88, 74.2, and 75 baud exist. The frequency shift between the two tones used to be standardised on 170Hz but now people are increasingly using 200Hz, especially with many of the new digital modes. As the difference between the two standards is relatively small there is no incompatibility between them.

As the actual data rate is relatively low, the bandwidths required are less than those needed for SSB. It is easy to gain a rough estimate of the bandwidth required by doubling the baud rate and adding the frequency shift. For example, a 50 baud transmission using a 200Hz frequency shift would require a bandwidth of 300Hz, so a 500Hz filter would be quite acceptable.

In view of the relatively narrow bandwidths and the need to be able to tune in the tones to approximately the required frequency, a receiver (or transceiver) with a reasonably slow tuning rate is needed. Typically it should be possible to tune to within about 50Hz.

One of the major problems with RTTY is that any interference causes the received data to be corrupted. Even under relatively good conditions it is very difficult to have a totally correct copy. When interference levels rise, as they normally do at HF, then copy can be very difficult. To overcome these problems, new data modes are now widely used that utilise the power of computers to detect and correct errors.

RTTY is still widely used on the HF bands, and with a data rate of around six characters a second this makes it ideal for 'live' chats as most people's typing is not too far off this speed.

AMTOR

To overcome the problems with RTTY a system known as *AMTOR* has been developed. It is popular on the HF bands because it gives more reliable communication, especially when interference is present. It achieves this by using a coding system that allows errors to be detected and corrected.

The system uses the same basic five-bit code as RTTY, but sent at a data rate of 100 baud. A total of seven bits are sent. The additional two bits are used to ensure that the transmitted data pattern always contains four mark bits and three space bits. From a knowledge of this expected pattern the receiver is able to detect an error and action can be taken to correct it.

In operation the transmitter sends out three characters. The receiver checks them to ensure they are correct. If they are, then an acknowledgement is sent back to indicate this. Then the next block of three characters can be sent out. If they have not been received correctly then this is indicated and the block is re-sent.

A block takes a total of 450 milliseconds (450ms) to send. Each character takes 70ms, giving a total of 210ms for the transmission. Then there is a window of 240ms for an acknowledgement to be received. This amount of time is allowed to take account of the delays that occur.

The method of sending where such an *automatic request for repeat* (ARQ) is

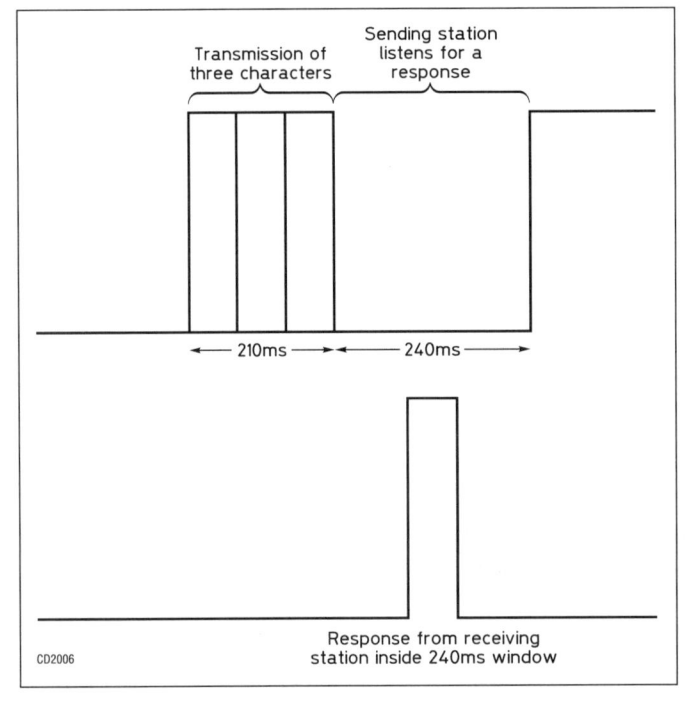

Fig 3.7. Timing of an AMTOR transmission in Mode A

used is called *Mode A*. However, it can only operate if contact has been established with a particular station. A general transmission such as a news bulletin or an amateur radio operator wanting a contact cannot work in this mode. Here a second mode called *Mode B* is used, where each character is sent twice. Initially the first character is sent once and then the repeat message is sent five characters behind the first one. The time interval between the two signals reduces the possibility of interference causing problems. Sending the data twice also gives the receiver two attempts at capturing each character. Also, because seven bits of data are sent instead of the five used for the character code itself, error detection is still possible, allowing the receiving equipment to decide which character of the two to accept.

Like RTTY the data rate is around five or six characters a second, dependent upon the level of interference, and as a result many people use this mode for chatting. There is also a certain amount of DX to be contacted, and indeed many people look for new countries and rare stations in the same way as they do on Morse and SSB, enabling this mode to cover a variety of interests.

PSK31

This data mode is gaining popularity and derives its name from the fact that it uses phase shift keying rather than frequency shift keying, and it transmits data at a rate of 31 baud. The mode is aimed at providing a greater level of performance for keyboard-to-keyboard, conversational-style data communications than is available with other data modes.

The aim is to provide an efficient yet straightforward system which does not use the complicated ARQ processes, and with only enough error correction to match the typical error rates that are encountered. Also, by using phase shift keying and a low data rate, it is possible to narrow the

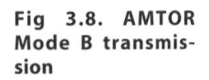

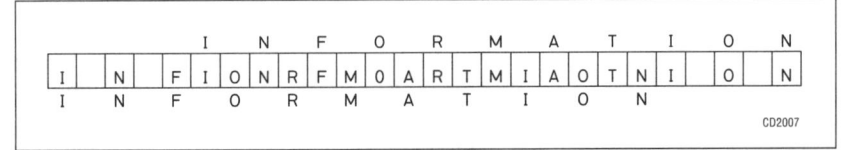

Fig 3.8. AMTOR Mode B transmission

bandwidth, which considerably reduces the effects of interference and noise. Bandwidths of 31Hz can be used, making this an extremely narrow-band mode, and one capable of operating under severe conditions.

Phase shift keying is rather different to the frequency shift keying that is widely used on the amateur bands today. It involves reversing the polarity, or phase, of the signal (180° phase shifts), and has been likened to reversing or swapping over the two wires in an antenna connection. However, in reality the phase reversals are not achieved in this way; instead they are generated and detected in the audio sections of the SSB transceiver being used.

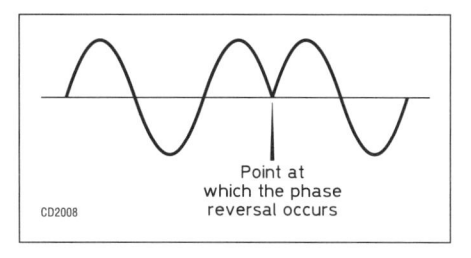

Fig 3.9. A phase reversal in a signal

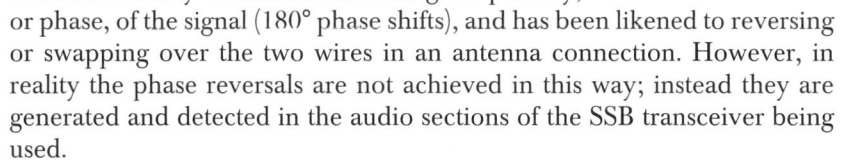

This form of modulation is known as *binary phase shift keying* (BPSK) and it is more efficient than either frequency shift keying, which has a greater bandwidth, or on/off keying which does not use the power as efficiently.

A novel form of data encoding is used. When sending asynchronous ASCII data, systems use a fixed number of data bits as well as start and stop bits. However, when sending a long run of data it is possible for the receiver to lose synchronisation. Additionally, improvements in speed can be gained from adopting variable-length codes with those codes that are used most often being the shortest. This is used to good advantage in Morse where the character 'e' (which is the most common in English) is a single dot. By analysing the occurrence of different ASCII characters a code called *Varicode* was devised. The shortest code, '00', was allocated to the space between two words.

It is possible to add error correction to the system. However, to achieve this it is necessary to use a form of keying called *quadrature phase shift keying* (QPSK). Instead of two phase states 180° from one another, QPSK uses four phase states, each 90° from one another. However, in operation on the bands it has been found that error correction with the use of QPSK only sometimes gives improvements over ordinary BPSK. Accordingly it is possible to use either system, dependent upon the conditions.

Further information about this interesting mode can be found in reference [1] or by visiting www.psk31.com.

Slow-scan television

Slow-scan television (SSTV) is used for a number of applications on the HF bands. For example, it is commonly used by radio amateurs to send pictures across the globe, and this adds a completely new dimension to making a contact. Many transmit pictures of themselves or their station.

SSTV uses the many of the basic principles of ordinary broadcast television but such transmissions occupy bandwidths of several megahertz, and obviously this cannot be accommodated on the HF bands. To overcome the problem the rate at which the picture is transmitted is slowed down so it can easily be accommodated within the 3kHz bandwidth occupied by a single sideband transmission. Actually, the required transmission bandwidth

A picture transmitted by slow-scan television

is between 1.0 and 2.5kHz. However, this does mean that the pictures transmitted by SSTV are normally still pictures.

To create an SSTV signal a picture is scanned in basically the same way as a normal television signal. A point is moved across the picture and the light level continuously detected. Once the point has completely moved across the picture it quickly returns to the original side of the picture, but slightly lower, and then it starts again. In this way the whole of the picture is scanned. At the receiving end the light levels detected at the transmitter are used to build up the original picture. The scanning directions are left to right for the line scan and top to bottom for the frame. Generally there are 120 or 128 lines per frame and the aspect ratio, ie the width-to-height ratio, is 1:1.

In order to ensure that the receiver and transmitter are synchronised, pulses are placed between each line and each complete picture or frame. Line synchronisation pulses are 5 milliseconds (5ms) long, whereas those for synchronising the whole picture or frame are 30ms in length. The picture information has a 60ms duration.

Fig 3.10. A slow-scan television video signal

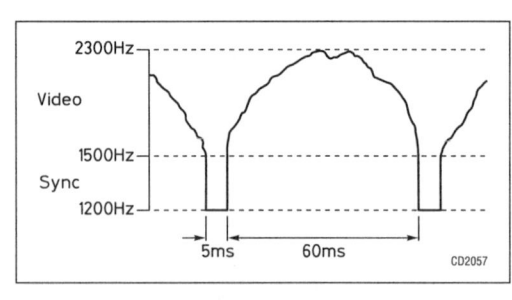

An example of a typical video signal is shown in Fig 3.10, and this is used to modulate the carrier. To achieve this an audio tone is varied in frequency in line with the light level synchronisation pulse. A frequency of 1200Hz is used for a frame pulse, 1500Hz for black and up to 2300Hz for peak white. This is used to give a single sideband signal for transmission. In essence this gives a radio frequency signal which varies by +400Hz for peak white, −400Hz for black, and −700Hz for a synchronisation pulse. As the synchronisation pulses are represented by a frequency lower than the one representing the black level, they are said to be *blacker than black*, and they cannot be seen on the screen.

Picture quality can vary widely. To ensure that the optimum quality is achieved it is essential to have good detection of the synchronisation pulses. Often a special filter is used to detect these 1200Hz pulses, even though they can easily be seen in the demodulated video signal.

There is a wide variety of different standards for picture size. Typically pictures are 128 lines and take eight seconds to send. Another standard is 256 lines, although pictures can be almost any length, terminated by a frame synchronisation pulse.

When slow-scan television first became popular, monitors containing tubes with a long persistence were required to display the pictures. Now most people use PCs running the relevant software to display the images and store them if needed. There are several packages that can be used, making the PC approach far more convenient and cost effective. In this system the audio is extracted from the headphone jack, or a line output socket on the receiver, and applied to a sound card in the PC.

Using computer software is also the most convenient way in which pictures can be generated. Often digital cameras may be used to take the photo and this information is downloaded into the computer for processing to produce the required graphic image. As in the case of the receiver, SSTV software packages enable the required audio to be generated from the graphic and this can be applied to the audio input of the transmitter for transmission.

Reference

[1] *RSGB Technical Compendium* (*RadCom* 1999), RSGB.

Further reading

Radio Communication Handbook, 7th edn, ed Dick Biddulph, M0CGN, RSGB, 1999.

CHAPTER 4

Receivers

In this chapter:

- Basic receiver concepts
- Direct conversion receiver
- Superhet receiver
- Frequency synthesisers
- Receiver specifications

THE performance of a receiver is of the utmost importance in any amateur radio station that is to be used on the HF bands, particularly when it is be used to pick out weak stations in the presence of high levels of interference. Any receiver will be able to receive some stations but the leading DXers will choose sets that are often much more expensive, proving their worth in being able to perform very well under exacting conditions. A good receiver, whether it is a separate item or included as part of a transceiver, is a key element in any station. However, choosing the right set is not always easy. There are plenty of different specifications to be appreciated in order to be able to compare one receiver with the next.

A typical HF receiver

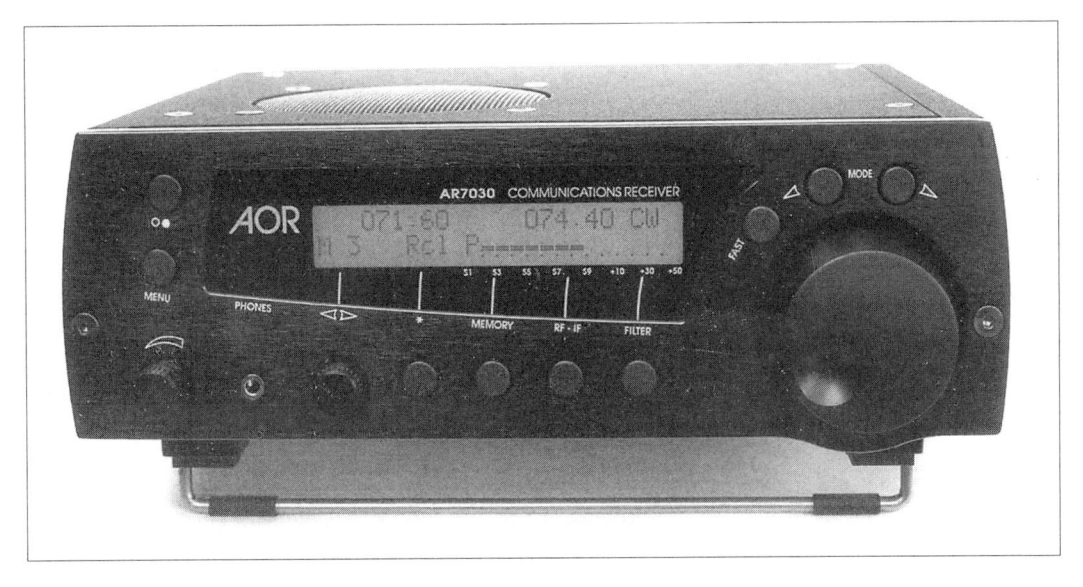

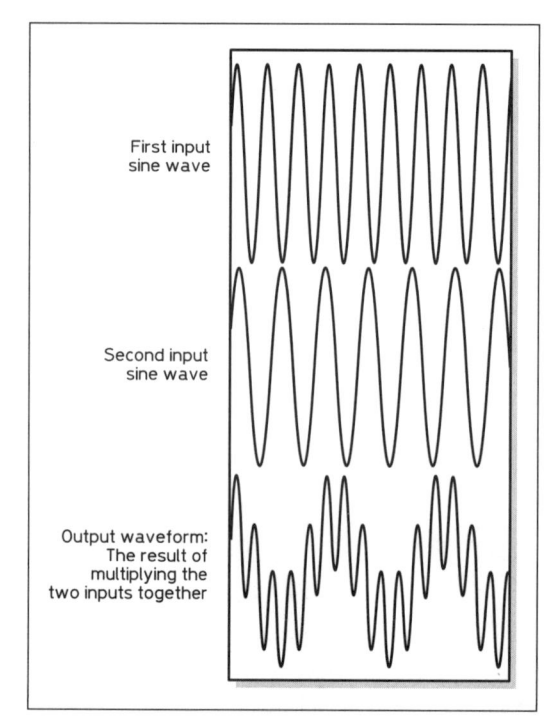

Fig 4.1. The output from a mixer as seen on an oscilloscope

Also, knowing how the different types of receiver work not only helps in choosing the right one, but also in gaining the best performance when it is being operated.

Receiver types

There are many HF receivers available on the market today. Not only are their specifications different, but there are two distinct types that are used. The first is the *direct-conversion receiver* where signals are mixed with an oscillator to give an audio output directly. The second main type is the *superhet*, where the incoming signal is mixed with a local oscillator to produce an output at an intermediate frequency. This is filtered and then the signal is demodulated to give the required audio. Both of these types of receiver rely on the principle of mixing for their operation.

Mixing

The mixing process is not an additive process like that in an audio mixer. Instead it is a multiplication process where the instantaneous output from the mixer is the product of the two inputs at that instant. A typical waveform is like that shown in Fig 4.1. If the output is analysed then it will be seen that the output contains signals with two new signals at the sum and difference frequencies of the input signals.

To look at how this works, if the input signals are at frequencies of f_1 and f_2, then new signals are produced at frequencies of $(f_1 + f_2)$ and $(f_1 - f_2)$. For example, if the input frequencies are 1MHz and 100kHz then new signals will be generated at 900kHz (difference frequency) and 1.1MHz (sum frequency).

Direct-conversion receiver

This type of receiver can be relatively simple but it can still provide an excellent level of performance. As a result it is very popular, and especially with low-power enthusiasts who build much of their own equipment. However, it does have some limitations, not being able to offer all the flexibility and facilities of a superhet receiver, and accordingly it is not generally used in high-performance sets.

The basic block diagram for this type of receiver is shown in Fig 4.3. Here the signals from the antenna enter the mixer along with a signal from

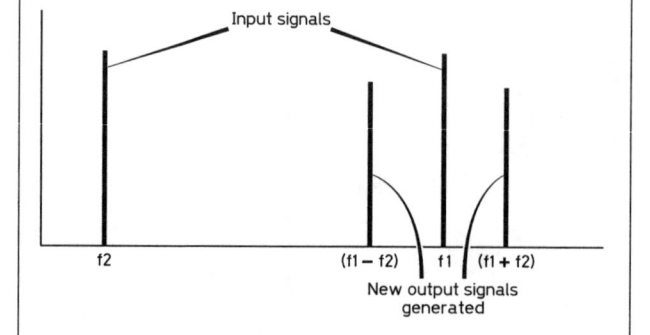

Fig 4.2. Signals produced by a mixer

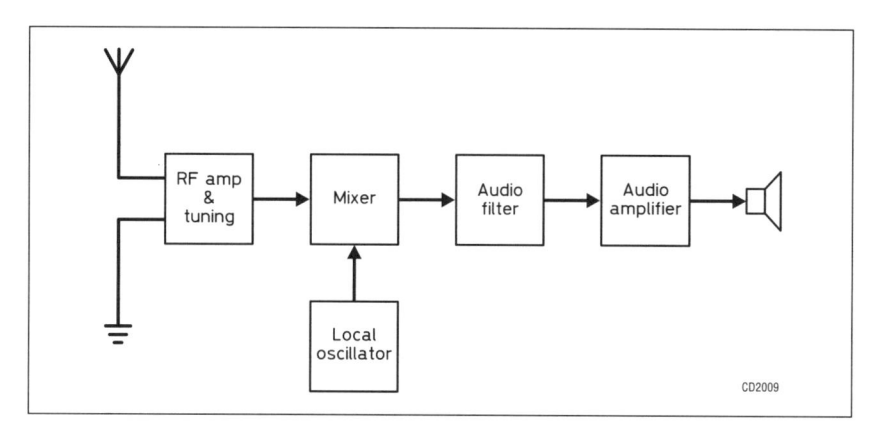

Fig 4.3. Block diagram of a direct-conversion receiver

a variable-frequency oscillator. The signals are mixed together, and the signals from the antenna that are close in frequency to that of the local oscillator produce audible 'beat note' signals. These pass through the audio low-pass filter and can be amplified to be heard in a loudspeaker. Those signals that are not close in frequency will produce mix signals that fall above the audio band and will not pass beyond the audio frequency low-pass filter.

To give an example, if the local oscillator is set to a frequency of 2.000MHz, then an incoming signal on a frequency of 2.001MHz will produce mix products at 0.001MHz (ie 1kHz) and 4.001MHz. The signal at 4.001MHz is well above the pass band of an audio low-pass filter and accordingly it is removed. The one at 1kHz is within the audio bandwidth and will pass through the filter to be amplified and fed into the loudspeaker. Another incoming signal on the slightly higher frequency of 2.040MHz will produce signals at the output of the mixer at frequencies of 4.040 and 0.040MHz. Both of these will be above the cut-off point of the low-pass filter and will not be heard. For communications purposes the cut-off frequency of the audio filter would be set to a frequency of around 3kHz, and in this way signals up to 3kHz either side of the local oscillator will be heard. Signals on different frequencies can be heard by varying the frequency of the local oscillator.

Some radio frequency tuning is normally placed prior to the mixer. This limits the signals entering the mixer to those in the region of the frequency of interest. Although the tuning is not required to be very sharp it helps prevent the mixer from becoming overloaded by having signals from the whole radio spectrum appearing at its input.

The major disadvantage of the direct-conversion receiver is that it operates by creating a beat note with the incoming signal that appears within the audio bandwidth. This is exactly what is required for Morse (CW) and SSB, but it is not ideal for AM where the receiver has to be tuned to be zero beat with the carrier. It is also not possible to resolve FM.

A further disadvantage is what is known as the *audio image*. This occurs because signals either side of the local oscillator appear within the audio band and can be heard. Using the example figures above, signals at both

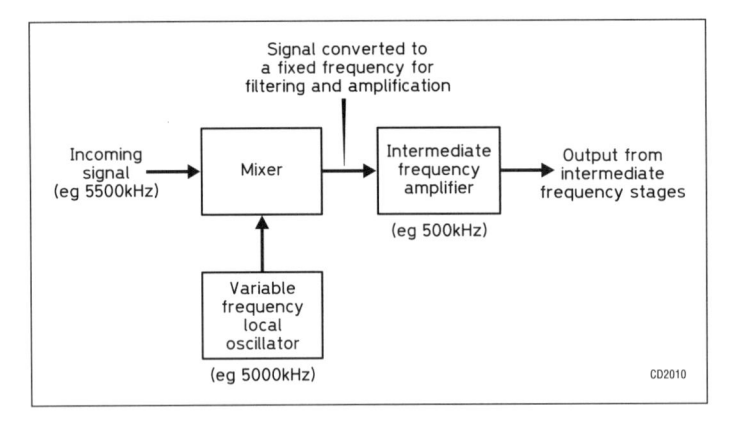

2.001 and 1.999MHz will produce beat note signals of 1kHz. This means that levels of interference are higher than for an equivalent superhet where this problem is not experienced.

Superhet receiver

The superhet is the most widely used form of receiver. Invented by Edwin Armstrong in 1918, it provides the flexibility and performance required by most receivers today. It operates by mixing the incoming signals with a locally generated oscillator signal to convert them down to a fixed *intermediate frequency* (IF). It is within the IF stages of a receiver that the selectivity is provided which enables the unwanted off-channel signals to be rejected. By changing the frequency of the local oscillator, signals on different frequencies can be converted down to the IF and hence the receiver can be tuned.

One problem with the system is that it is possible for signals on two different frequencies to enter the IF. Take the example of a receiver with an IF of 500kHz and a local oscillator tuned to 5MHz. A signal with a frequency of 5.5MHz will mix with the local oscillator to give an output at 500kHz. This will pass through the IF filters. It is also possible for a signal on 4.5MHz to mix with the local oscillator to give an output at 500kHz. To prevent this happening, a tuning circuit is required before the mixer so that only the required frequency band is allowed through as shown in Fig 4.5. The unwanted signal is known as the *image*.

Looking at the overall block diagram of the receiver it can be seen that the signal first enters the RF amplifier stage. The purpose of this stage is two-fold: to provide sufficient selectivity to prevent image signals reaching the IF stages, and to provide signal amplification before the mixer.

From the RF amplifier the signal passes to the mixer where it is converted down to the intermediate frequency. While the block diagram only shows one conversion, this process may be undertaken in several stages. The reason for this is that if a high intermediate frequency is chosen, the image frequency will be further away from the wanted one, enabling the image response to be improved. A high IF will also mean that that achieving the required level of adjacent channel selectivity is more difficult, whereas a low IF will mean the image rejection is

Fig 4.4. The superhet principle

Fig 4.5. Front-end tuning to remove the image signal

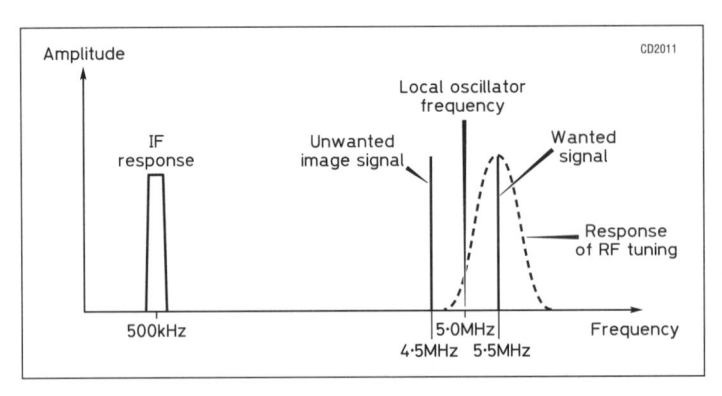

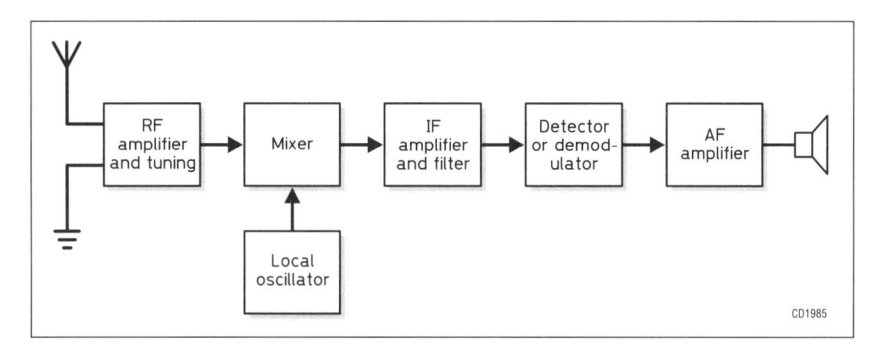

Fig 4.6. Block diagram of a basic superhet receiver using a single conversion. Some sets have two or three conversions to give the required level of performance as described in the text

poor. Many receivers adopt a two- or even a three-stage conversion process to enable the required balance between all the conflicting factors to be achieved.

Often HF receivers may actually convert the incoming signal up in frequency at the first conversion to achieve the required image rejection. Afterwards it is converted down in frequency to enable the required degree of adjacent channel selectivity to be achieved.

The variable-frequency local oscillator is important, and a variety of approaches can be used. The most straightforward is to employ a free-running, LC-tuned oscillator (ie an oscillator using a capacitor and inductor for the resonant circuit that determines the frequency of its operation). These were widely used up until the 'seventies but had the drawback that they tended to drift in frequency. Nowadays frequency synthesisers are used, and these circuits offer many advantages in terms of stability and flexibility of operation. They are described later.

The IF circuits in a receiver provide the majority of amplification and selectivity. In receivers using discrete components several stages of amplification are used but often now just one integrated circuit is likely to be employed to give the required level of gain. As the IF provides the adjacent-channel selectivity for the receiver, the filter used in these stages governs the performance of the whole set. Accordingly a high-performance filter is used which in most sets is a crystal type.

The output from the IF amplifier is connected to the demodulator. As different types of demodulator are needed for different types of transmission there may be two or more different circuits that can be switched in. For single sideband or CW, product detectors or demodulators with a beat frequency oscillator (BFO) are used. Essentially the product detector is a mixer and the BFO beats with the incoming signal to give an audio note in the case of CW or regenerates the audio signals in the case of single sideband.

Once the signal has been demodulated it can be amplified and connected to a loudspeaker, headphone or, in the case of data transmissions, it can be connected to the data equipment.

Frequency synthesisers

Frequency synthesisers have already been mentioned and they are used almost universally for local oscillators in receivers and transceivers today.

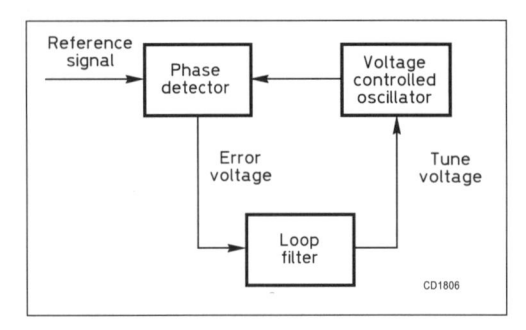

Fig 4.7. A basic phase-locked loop

Many scanners or receivers may also boast terms like 'PLL' or 'synthesised' because they include frequency synthesisers. They offer many advantages over other techniques that could be used. For example, they can be very easily controlled by microprocessors, their frequency stability is excellent and they are exceedingly versatile, allowing many facilities to be incorporated into sets, including scanning, multiple VFOs and memories.

Although synthesisers can take many forms, the type which has gained universal acceptance is based around the *phase-locked loop* (PLL). A basic loop is shown in Fig 4.7 and from this it can be seen that it consists of a phase detector, voltage-controlled oscillator (VCO), a loop filter and finally a reference oscillator. As the frequency stability of the loop or synthesiser is totally governed by the reference oscillator this is crystal controlled, and in professional equipment it often uses a temperature-controlled oven.

The operation of the basic phase-locked loop is fairly straightforward. The reference oscillator and VCO produce signals that enter the phase detector. Here an error signal is produced as a result of the phase difference between the two signals. This signal or voltage is then passed into a filter that serves several functions. It controls the loop stability, defines many of the loop characteristics and also reduces the effect of any sidebands that might be caused by any of the reference signals appearing at the VCO input.

Once through the filter the error voltage is applied to the control input of the VCO so that the phase difference between the VCO and reference is reduced. When the loop has settled and is locked, the error voltages will be steady and proportional to the phase difference between the reference and VCO signals. As the phase between the two signals is not changing, the frequency of the VCO is *exactly* the same as the reference.

In order to use a phase-locked loop as a synthesiser, a divider is placed in the loop between the VCO and phase detector. This has the effect of raising the VCO signal in proportion to the division ratio. Take the example when the divider is set to 2. The loop will reduce the phase difference between the two signals at the input of the phase detector, ie the frequency of both at the two phase detector inputs will be the same. The only way this can be true is if the VCO runs at twice the reference frequency. Similarly if the division ratio is 3, then the VCO will run at three times the reference frequency. In fact by making the divider programmable the output frequency can be easily changed.

In order to have channel spacing of 25kHz or less the reference frequency has to be made very low. This is generally done by running the reference oscillator at a relatively high frequency, eg 1MHz, and then dividing it down as shown in Fig 4.8.

This is the basic synthesiser loop that is at the heart of virtually all frequency synthesisers. It can be enhanced in many ways to give more flexible operation and smaller step sizes that give almost continuous tuning.

Fig 4.8. A basic frequency synthesiser loop

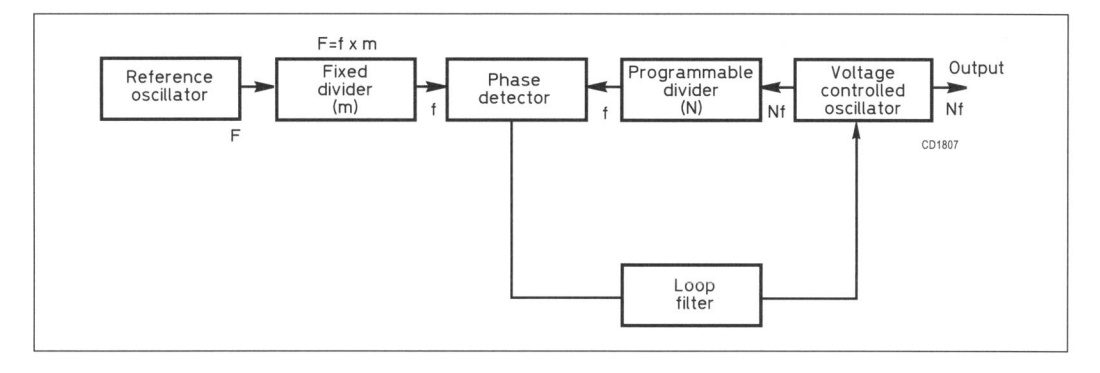

Selectivity

The selectivity of a receiver governs the way in which it accepts signals on the wanted frequency and rejects those that are off-channel. The most obvious way in which this affects a receiver is in terms of its adjacent channel selectivity, although there are other areas of the receiver where selectivity is important.

The adjacent-channel selectivity of a receiver is governed by the performance of the IF filter. Its performance will determine how well the whole receiver operates in terms of rejecting signals that are just off-channel. It is worth noting that for different types of emission different bandwidths are required. For Morse transmissions bandwidths of 250 or 500Hz are mostly used, while for SSB figures of between 2.5 and 3kHz are common.

In an ideal world the response of a filter would be like that shown in Fig 4.9(a). However, in reality it will look more like that shown in Fig 4.9(b). By comparing these it can be seen that there is a significant difference.

The two main specifications of importance are the bandwidths of the pass band and the stop band. The bandwidth of the pass band is defined as being the bandwidth between the two points where the voltage response drops by 6dB (ie to a quarter of the power level) from its in-band level. This is the figure quoted as the bandwidth of the filter above and is taken as the range of frequencies that are accepted by the filter. The stop band is equally important. It is normally the bandwidth where the response has fallen to a level where it should not let signals through. This is normally taken as the point where the response has fallen by 60dB from its in-band level.

The difference in bandwidth between the pass band and stop band shows how quickly the response falls away. Ideally they should be almost the same but this can never be the case in reality. To give an indication of the shape of the

Fig 4.9. Response of ideal and real filters

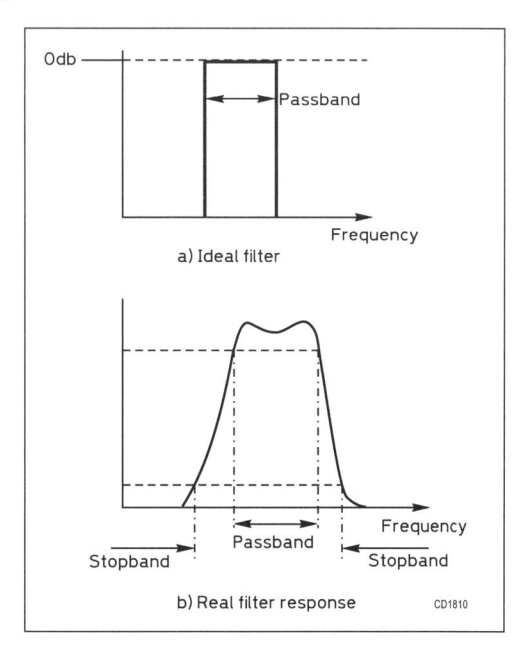

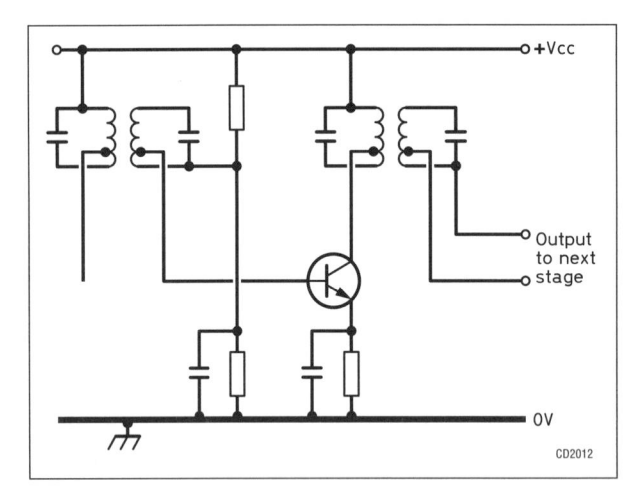

Fig 4.10. A typical IF stage using a transformer and discrete components

filter response a figure known as the *shape factor* is sometimes quoted. This is simply the ratio between the pass band and the stop band bandwidths. Thus a filter having a bandwidth of 3kHz at −6dB and 6kHz at −60dB would have a shape factor of 2:1. For this figure to have real meaning the two attenuation figures must also be quoted. For example, a specification might quote a shape factor of 2:1 at 6/60dB.

There are several different types of filter that can be used in a receiver. The simplest types are LC filters, and in many older sets the interstage coupling transformers are tuned to give the selectivity. A typical circuit showing how this is accomplished is shown in Fig 4.10. Normally one transformer is found on the output of the mixer, one for each interstage coupling that is required and one on the output of the IF strip prior to the demodulator. In broadcast sets, for example, three transformers may be found. The problem with these LC filters is that they do not give the very high levels of selectivity required on today's amateur bands.

Crystal filters are normally used to achieve the very high levels of selectivity required by amateur receivers and transceivers. These use the *piezo-electric effect* for their operation, which means that an electrical signal entering the crystal will cause sympathetic mechanical vibrations to be set up in the latter. Similarly mechanical vibrations on the crystal will give rise to an equivalent electrical signal. In this way the electrical circuit may use the mechanical resonances of the crystal, and as these possess very high degrees of selectivity or Q, they are able to be used to produce very-high-performance filters. Normally several crystals are used to enable the required performance to be obtained.

In some cases filter specifications for a receiver or transceiver may include a figure for the number of *poles* in the filter. To explain this in any detail requires looking at some filter theory. However, it is sufficient to say that there is a pole for every crystal in the filter. For example, an eight-pole filter has eight crystals. In fact most filters today have six or eight poles.

While most of the filtering is performed in the IF stages of a superhet receiver, sometimes filtering in the audio stages can be used to very good effect. Normally audio filters are designed around operational amplifiers using only capacitors and resistors. With these circuits, very high degrees of selectivity can be achieved without the use of expensive components. Often these filters are used in the reception of Morse signals where very narrow bandwidths can be used. The main disadvantage of an audio filter is that the AGC is derived from the signal appearing at the output of the IF. This can be a disadvantage if there is a strong signal within the IF pass band but outside the pass band of the audio filter. In cases like these the strong signal will 'capture' the automatic gain control (AGC), altering the gain level of the receiver. A weak signal within the pass band of the audio filter will then be altered in strength by the AGC level variations determined by the strong signal.

Image response

The image response of a receiver is very important. If the performance of the receiver is poor in this respect, unwanted signals will appear and cause interference, possibly masking out the wanted signals. The levels of performance can be particularly critical if weak signals are being received on a quiet band, and the image frequency falls on a band where there are strong signals, for example from broadcast stations using many hundreds of kilowatts.

Image response rejection figures are normally expressed as a certain number of decibels rejection at a given frequency. For example the image rejection may be 60dB at 30MHz. This means that a signal on the image frequency must be 60dB greater in strength than one on the wanted channel to produce an output of the same level. Although this may appear to be a large amount, signals picked up by receivers vary in strength by enormous amounts and often more than 60dB.

It is found that image rejection varies with the frequency of the receiver, degrading at the higher frequencies. The reason for this is that the difference between the wanted signal and the image is a smaller proportion of the operating frequency. As a result the frequency at which the image response specification is made is quoted.

Noise performance and sensistivity

The noise performance of a receiver is very important. If it introduces any noise onto the signal, then it will tend to mask it out, reducing readability or possibly completely obliterating it in extreme cases. The noise performance of the first stages in the receiver are the most important. As each successive stage within the receiver will amplify noise from the previous stages, this means that the design of the first RF amplifier is very critical and it should be optimised to give the lowest noise.

While the noise performance of the receiver is important, it is not as crucial as it is for sets used on the VHF and UHF bands. The reason for this is that the level of noise picked up by the antenna on the HF bands generally exceeds that generated in the receiver. This noise comes from a variety of sources (Fig 4.11). Some is man made, and this is obviously lower in rural areas, and tends to be highest in business or industrial areas. Another constituent of the received noise is atmospheric noise. The level of this is dependent upon a variety of factors, including the time of day, atmospheric conditions and a variety of other factors. Some galactic noise may also be present.

The simplest method of quantifying the noise performance of a receiver is to quote the signal-to-noise ratio for a given input signal. The measurement

Fig 4.11. Noise field strength values in a 2.7kHz bandwidth

of the difference between the signal and the noise indicates the noise performance and hence the sensitivity of the receiver (Fig 4.12).

For this measurement to be meaningful the bandwidth of the receiver must also be given. The reason for this is that the wider the bandwidth, the greater the level of noise that is received. When used with amplitude-modulated signals the depth of modulation must also be given

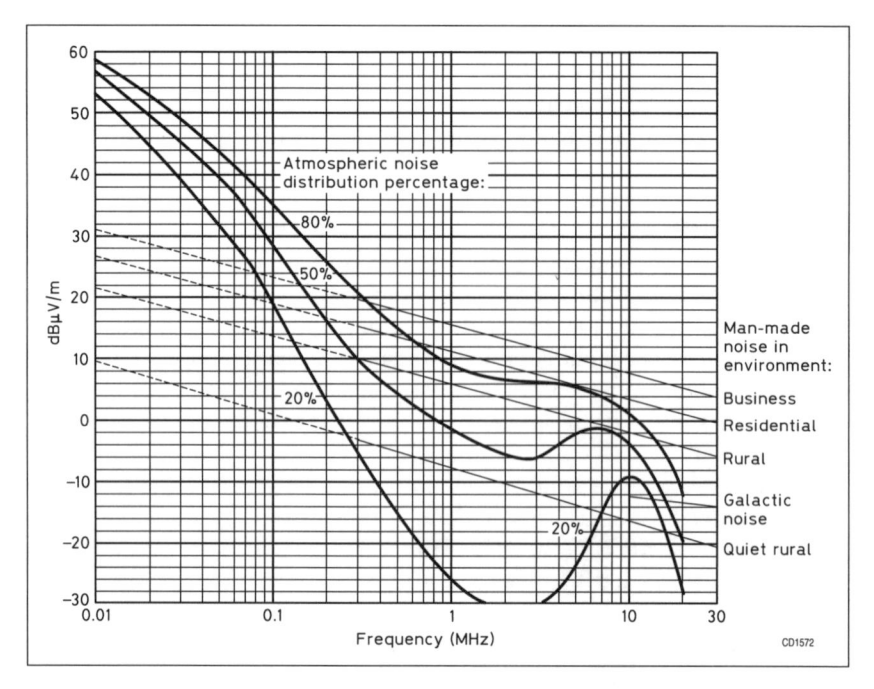

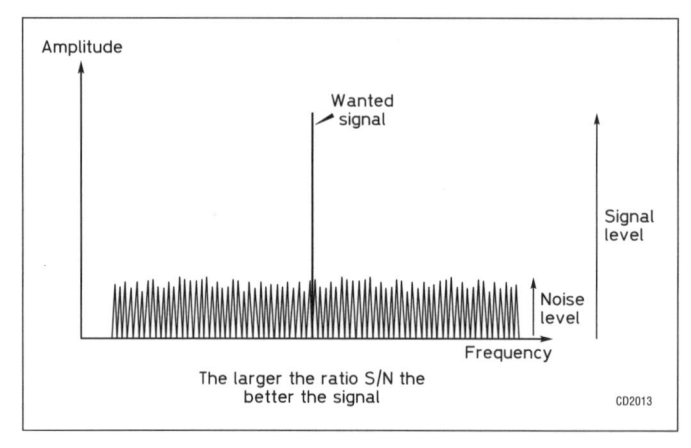

Fig 4.12. Concept of signal-to-noise ratio

because this will vary the level of the output signal from the loudspeaker. This method is widely used for HF receivers, and here a figure of around 1 microvolt for a 10dB signal-to-noise ratio in a 3kHz bandwidth for SSB or Morse reception is common. For AM a figure of 1.5 microvolts for a 10dB signal-to-noise ratio in a 6kHz bandwidth at 30% modulation may be achieved.

Strong-signal handling performance

Although it is important for any receiver to be sensitive, it is equally important for it to be able to withstand strong signals without overloading. If it cannot handle strong signals well, then the more interesting ones that are often quite weak may be masked out by the effects of overloading.

There are a number of parameters that relate to the strong-signal handling capability of a receiver. Two of the most important ones are the third-order intercept and the third-order intermodulation distortion specifications.

Receiver amplifiers do not have an infinite signal-level range. The output will follow a linear relationship for input signal levels up to a certain point. Beyond this point the amplifier is unable to provide the required output level and the signal becomes compressed. Unfortunately the non-linearity brings a number of problems with it, including intermodulation products, cross-modulation, and blocking.

A number of problems arise when an amplifier runs into non-linearity. Harmonics are generated and signals mix with one another. These products on their own do not normally cause a problem as the front-end selectivity means that signals that could enter the amplifier generate signals that are outside the range of interest. Take the example of a receiver set to around 10MHz. If there is a strong signal at 10.1MHz then the harmonics will fall at 20.2, 30.3MHz and so forth. These are well outside the range of the receiver and they are clearly going to have no effect on reception. Similarly signals that could enter the amplifier and mix with one another also fall outside the range of interest. Take the example of signals at 10.0MHz and 10.1MHz. They will mix to produce signals at 0.1 and 20.1MHz, and these are clearly not going to affect reception.

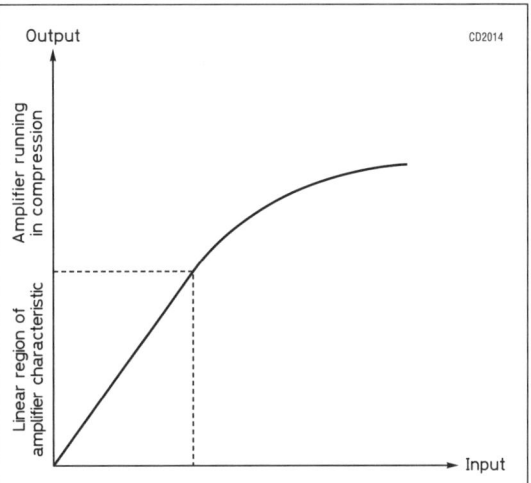

Fig 4.13. Compression in an amplifier results in the output being compressed beyond a certain input level, giving rise to problems such as intermodulation, cross-modulation and blocking

However, problems arise when a harmonic of one signal mixes with the fundamental of another, eg $(2f_1 - f_2)$, this being known as a *third-order product*. The third harmonic of one may mix with the second harmonic of the other, eg $(3f_1 - 2f_2)$, a *fifth-order product*. To put figures against the formulae, take two signals at 10.0 and 10.1MHz. The third-order product is 9.9MHz, and the fifth-order product is 9.8MHz.

Similarly signals are produced at 10.2 and 10.3MHz. In fact, a 'comb' of signals stretches out from the two main signals, each separated by the frequency difference between them (see Fig 4.14). As these signals are in the same region as the incoming signals it is possible for them to pass into later stages of the receiver.

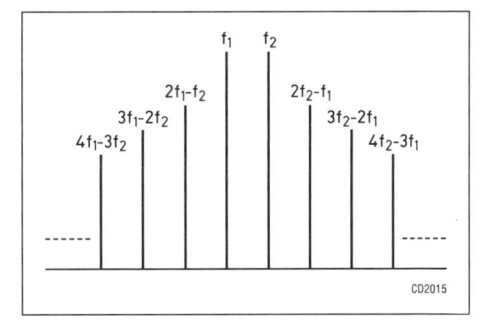

Fig 4.14. Intermodulation products stretch out either side of the two incoming signals. It is possible for two strong signals to pass through the RF tuning, causing the RF amplifier to run into compression and generating intermodulation products that can enter the following stages

Third-order intercept

To gain an idea of the resilience of a receiver to intermodulation distortion, a concept known as the *third-order intercept point* is used. The figure uses the fact that the third-order products rise in level much faster than the wanted ones as the input level rises.

A plot can be made of the output for varying input levels. As the input signal level is increased, the output level rises in line with it. Initially no third-order products are seen because they are masked out by the noise but, as the input level increases, they start to rise. As the input level is increased still further, the amplifier will start to 'limit' and, if the curves for the levels of the wanted signals and the third-order products are extrapolated or continued, it will seen that they cross. The point at which this occurs is known as the *third-order intercept point* and it is measured in watts – or in most cases milliwatts. The higher the level of the intercept point, the better the strong-signal handling of the receiver. Typically an amateur set might have an intercept point of 15dBm, ie 30 milliwatts (30mW).

Blocking

Sometimes when a very strong signal is received, it can 'block' the front-end of the receiver, reducing the level of other signals. This is a problem if it happens when a weak signal is being received, and can arise during contests when many very strong signals may be present on the band.

Blocking occurs because the receiver front-end runs into compression. When this happens it has the effect of reducing the level of the other signals passing through it in a similar way to the capture effect encountered in the reception of FM signals.

The amount of blocking is dependent upon the level of the incoming signal. Another factor is the frequency offset from the channel being monitored – the further away the strong signal is located, the less will be its effect because the selectivity, particularly in the front-end stages, will reduce its level. In view of these variables, blocking is specified as the level of the unwanted or strong signal which reduces the gain of the receiver by 3dB at a given offset which is normally 20kHz. A good receiver should be able to withstand signals of more than about 10mW before the gain is reduced by 3dB.

Cross-modulation

This is another effect that can be noticed on occasions, and was particularly important when amplitude-modulated signals were more popular. When it occurs it is found that the amplitude variations or modulation on a nearby strong signal is heard on weaker signals close by.

Cross-modulation is a third-order effect and it is found that a receiver with a good third-order intercept point is likely to have good cross-modulation performance.

To specify cross-modulation performance, the effect of a strong amplitude-modulated carrier with a known level of modulation is noted on a small wanted signal The specification involves several figures. The strength of the strong signal must be stated, along with the level of modulation, the frequency offset from the weak wanted signal and the level of the wanted signal. Generally the specification is worded as the level of a strong carrier with 30% modulation to produce an output 20dB below that produced by the wanted signal with a 20kHz offset. Usually the wanted signal level is taken as −53dBm (ie 53dB below 1mW), and into a 50-ohm load this works out to be 1mV.

Dynamic range
Both sensitivity and strong-signal handling capacity are important in any receiver. Accordingly the *dynamic range*, or range over which the receiver can operate, is very important. It is defined as the difference between the weakest signal that the receiver can receive and the strongest one it can tolerate without any noticeable degradation in performance. Unfortunately there are a number of ways in which the points at both ends of the scale can be measured and care needs to be taken when assessing dynamic range specifications.

The weakest signal that can be received is governed by the sensitivity. A term called the *minimum discernible signal* (MDS) is often used. It is generally taken as a signal equal in strength to the noise produced by the set, and its level is generally given in dBm, ie decibels relative to a milliwatt for a given receiver bandwidth. The bandwidth must be included because the level of noise is dependent upon this. Typically it might be around −135dBm for a bandwidth of 3kHz.

At the other end of the scale there are two main factors that come into play. One is the generation of intermodulation products, and the other is blocking. As the onset of these effects occurs at different signal levels, the way in which the high end of the dynamic range is determined must be mentioned in the specification.

Even when blocking is used to determine the high end of the specification, different manufacturers use different levels of blocking. Often a 1dB decrease in sensitivity is used, but sometimes 3dB is used instead.

Where the level of intermodulation products is used to determine the top end of the range, it is normally taken that the level of these products must not exceed the MDS, ie be no greater than the noise floor of the set.

Most modern sets have an intermodulation-limited dynamic range of between 80 and 95dB. If a specification using the blocking figure is taken instead, then a range of 115dB or possibly even more might be expected. The difference between figures obtained by the different methods highlights the care that must be taken when comparing sets from different manufacturers where different measurement methods may be used.

Phase noise and reciprocal mixing
Frequency synthesisers have many advantages but one of their main disadvantages is that some designs generate large amounts of *phase noise*. Some such noise is present on any oscillator signal, and essentially it manifests itself as a small amount of frequency variation or 'jitter' on the signal. If the signal is examined on a spectrum analyser, it is seen as noise spreading out either side of the wanted signal as shown in Fig 4.15.

Different types of oscillator produce different levels of phase noise. Crystal oscillators are generally the best, and LC-tuned variable frequency oscillators are normally very good. In general, the higher the Q (quality factor) of the tuned circuit, the lower the level of phase noise. However, the way in which frequency synthesisers operate means that some can produce high levels of phase noise.

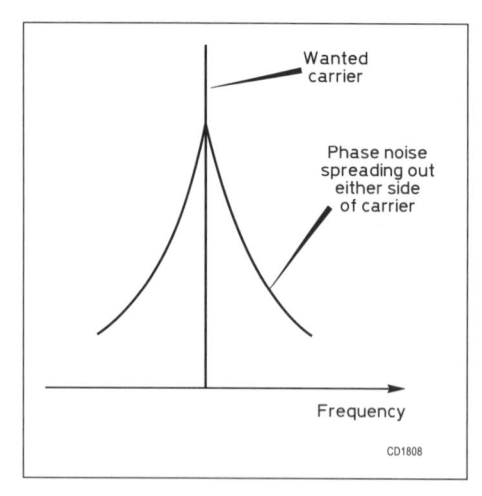

The main way in which phase noise affects receiver performance is through what is termed *reciprocal mixing*. To look at how this occurs, take the example of a superhet receiver tuned to a strong signal. The signal will pass through the radio frequency stages to the mixer where it will be converted down to the intermediate frequency stages, passing through the filters to the demodulator. When the receiver is tuned off frequency by 10kHz, for example, the signal will no longer be able to pass through the intermediate frequency filters. However, it will still be possible for phase noise on the local oscillator signal to mix with the incoming signal to produce noise that will fall inside the receiver pass band as shown in Fig 4.16. In some cases this might be strong enough to mask out a weak wanted signal.

A number of different methods are used to define the level of reciprocal mixing. Generally they involve measuring the response of the receiver to a large off-channel signal. These measurements are rarely easy as the performance of any signal generator must be very good – significantly better than the receiver – otherwise the performance of the signal generator is measured.

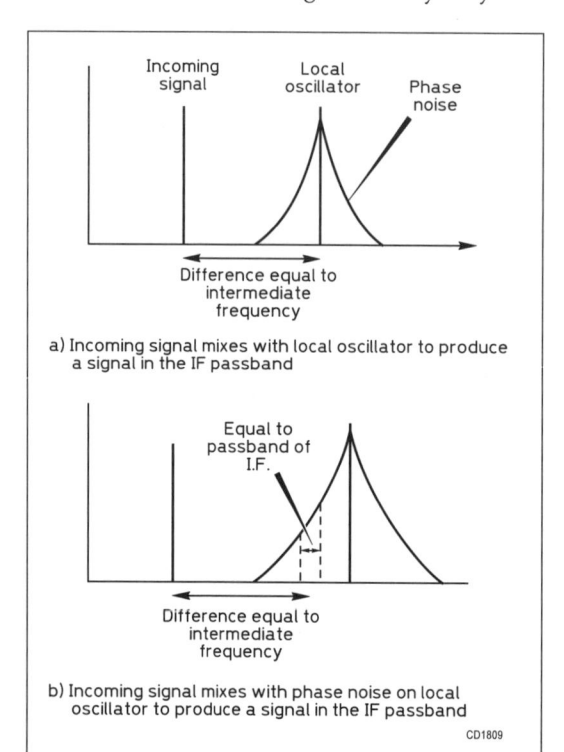

A measurement can be made by noting the level of audio from a small signal when the BFO is switched on. The signal is then tuned off-channel by a given amount, often around 20kHz, and the signal increased until the same audio level is achieved as a result of the receiver phase noise. As the level of noise is dependent upon the receiver bandwidth, this has to be stated in the specification as well. A good HF receiver might have a figure of 95dB at a 20kHz offset using a 2.7kHz bandwidth. This figure will naturally improve as the frequency offset is increased. At around 100kHz one might expect to see a figure in excess of 105dB or possibly more.

Another way of measuring the phase noise is to inject a large signal into the receiver and monitor the level needed to give a 3dB

Fig 4.15. A signal with phase noise

Fig 4.16. Reciprocal mixing

CHAPTER 4: RECEIVERS

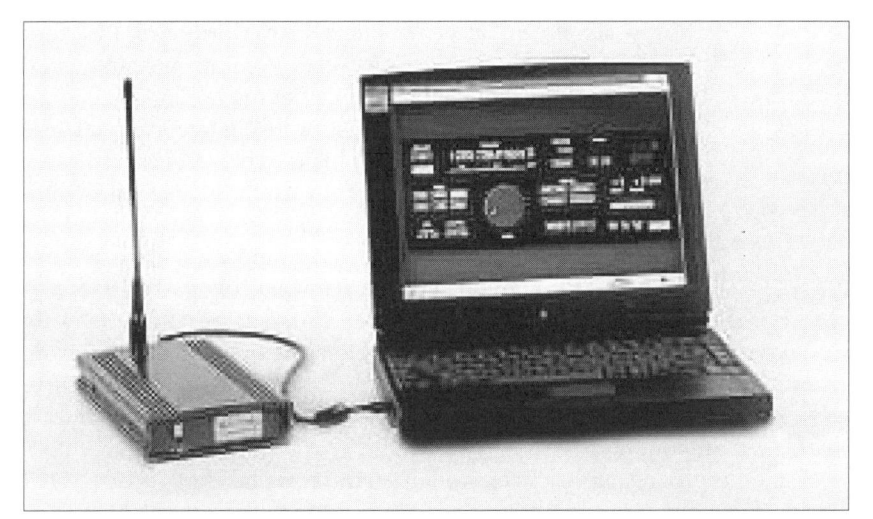

Winradio uses the processing power of a PC to perform the digital signal processing, control of functions like the tuning as well as providing the interface for the receiver controls *(photos courtesy Winradio)*

increase in the level of background noise. As a variety of measurements can be used, it is often best to study the reviews in the magazines made by the same person because he will use consistent measurement techniques. In this way several sets can be compared against one another.

Digital signal processing

Today many new pieces of equipment advertise the fact that they use *digital signal processing* (DSP) techniques. As the technology for DSP becomes cheaper it will be used increasingly in the years to come, giving the added advantage of improved performance.

DSP involves converting a signal into a digital format and then processing the signal using computer techniques. This processing is able to perform the actions that are traditionally performed by analogue components like filters, amplifiers, mixers and demodulators. As these functions are performed mathematically they do not have many of the limitations of the analogue components, and they are able to achieve much higher levels of performance. While this may seem like a dream come true, there are still limitations and it is not, for example, possible to produce the perfect filter. However, by eliminating the limitations of the analogue circuitry it is possible to achieve

47

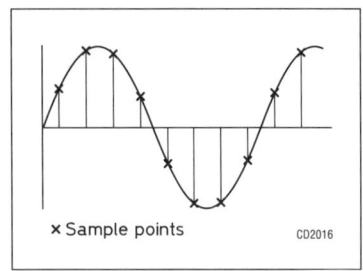

Fig 4.17. Converting a signal into a digital format by sampling at regular intervals of time.

very significant improvements in performance. Another advantage is that further functionality or improved filters can be added simply by changing the software.

The first step in the process is to change the analogue signal into a digital format. This is achieved using an *analogue-to-digital converter* (AD converter). The signal is sampled at regular intervals of time, and the voltage at that time is converted into a digital format as shown in Fig 4.17.

Once the signal is in digital format it is passed in to the signal processing section. The core of this section is generally a specially designed processor, optimised for digital signal processing. Here the sequential values representing the signal are processed mathematically to provide in a digital way all the functions that would normally be provided by analogue components.

Once the processing has been undertaken the signal is often converted back into an analogue format using a *digital-to-analogue converter* (DA converter) to be passed into an audio amplifier and then to headphones or a loudspeaker as required.

With the advances being made in technology DSP is becoming cheaper to implement, and as a result many more amateur HF (and other) receivers and transceivers are using DSP back-ends. Normally it is only used to replace the analogue IF and demodulator stages. However, there are also some external DSP filters that can be added to various sets.

CHAPTER 5

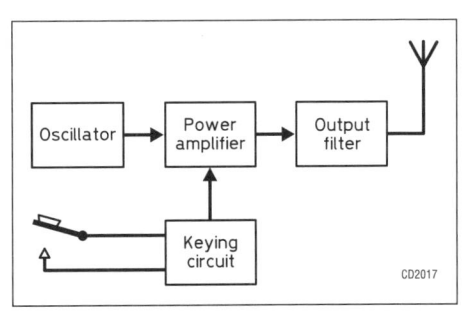

Transmitters

In this chapter:

- QRP transmitters
- High-power transmitters
- Linear amplifiers
- VOX and break-in
- Transmitter specifications
- Speech processing

THERE are many transmitters from which to choose these days. Often they are part of a transceiver because this gives the maximum amount of flexibility and ease of use. Even so, the way in which a transmitter operates and the many specifications that are relevant still apply in the same way as if it was a stand-alone unit. Like receivers, these specifications need to be understood when buying a unit to make sure it meets your requirements, and an appreciation of how a transmitter works makes it possible to use the equipment more effectively and reliably.

QRP-style transmitters

Today a large number of people like to venture onto the bands with low-power equipment they may have built themselves. There is a great feeling of achievement in making contacts with such equipment, especially if you have constructed it yourself. The transmitters used are often relatively simple, and generally only cater for Morse. In this way the complexity can be reduced to a level where they can be confidently tackled as a constructional project by most people.

A simple Morse transmitter can be made using as few as three transistors and a handful of other components. Typically this would be one transistor used as a crystal oscillator and the second as the power amplifier as shown in the block diagram of Fig 5.1. A third transistor is often used to switch the PA and prevent the key from having to pass the full PA current which might cause pitting or burning of the key contacts. While a crystal oscillator

Fig 5.1. Block diagram of a simple QRP CW transmitter

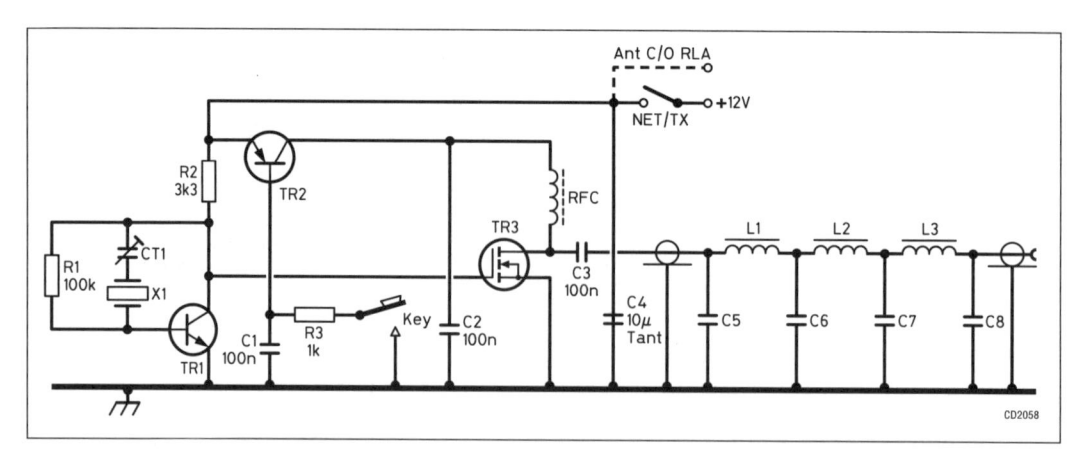

Fig 5.2. Circuit of a typical QRP transmitter

reduces the flexibility of the transmitter in terms of being able to set the frequency on the required spot, it simplifies the circuit considerably. A variable-frequency oscillator (VFO) requires the use of several more transistors and this complicates the construction. However, a VFO does enables the frequency of operation to be chosen. This can be a great advantage when running QRP as it means that other stations can be called rather than having to rely calling CQ, or hoping that other people just happen to be on the same frequency as your crystal.

On the output of the transmitter there is a filter and matching network. This provides two functions. The first is to provide a good match between the output transistor and the antenna, thereby ensuring that the maximum amount of power is transferred into the latter. The second function is to filter any out-of-band spurious products. Any amplifier, and particularly one running in Class C, will produce harmonics, and these need to be removed to ensure they do not cause interference. Typically they should be about 50dB below the carrier level.

There are many transmitter designs that appear in the magazines. Alternatively there is a growing number of companies that sell kits, which are usually easy to make and can often be completed in an evening or two. In many cases metalwork can also be bought with the kit, and this provides a

A kit QRP transmitter. These kits are often very easy to build and provide a very cost-effective way of putting a signal on the air

A typical HF transceiver

very neat case for the unit, giving it a professional look. If you are not keen on metalwork this can be an ideal solution.

Usually QRP transmitters run at powers of 10W and below. They can be great fun to build and use. However, the very fact that they run low levels of power means that contacts will not be as plentiful as they are for those who have higher-power transmitters. Accordingly time should be spent in making the antenna as efficient as possible.

Single sideband transmitters

While there is a large group of QRP operators and constructors, most people tend to use higher-power, multimode transmitters that are part of a transceiver. These normally transmit either Morse or single sideband, and some even have frequency-modulation capabilities.

The way in which the signals are generated is very much the inverse of that found in a superhet receiver. This means that most of the circuits used in the receiver can also be used in the transmitter section, saving the duplication of many areas of circuitry. However, in order to explain the way in which a transmitter works, the circuitry of one outside the confines of a transceiver will be described. Exactly the same principles hold for the transmitting sections of a transceiver.

The block diagram of a basic transmitter is shown in Fig 5.3 and it can be seen to be fairly straightforward. First, the carrier is generated using a crystal oscillator. Often a frequency in the region of 8 to 10.7MHz is chosen. This is high enough to simplify the problem of removing unwanted mix products later on when the sideband signal is converted to its final frequency. This is particularly important where the transmitter forms part of a transceiver and unwanted mix products may appear in the receiver as 'birdies'. Another advantage of using an IF in this frequency range is that it is not so high that suitable filters are prohibitively expensive.

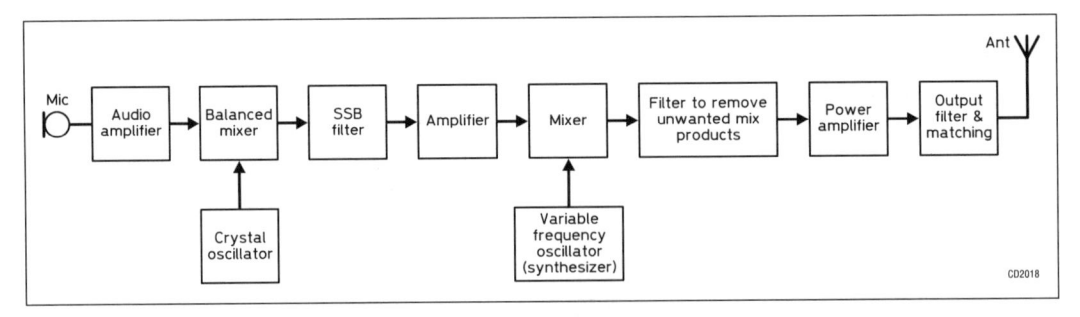

Fig 5.3. The block diagram of an elementary SSB transmitter

The signal from the carrier oscillator is applied to a balanced modulator, together with the amplifier and processed audio. The action of this modulator (mixer) produces a signal containing two sidebands (the sum and difference frequencies). A double-balanced mixer is normally used and this means that the input signals are suppressed at the output, giving a double sideband, suppressed carrier signal. This signal is then applied to a sideband filter to remove the unwanted sideband, leaving a single sideband, suppressed carrier signal at the intermediate frequency.

With the basic signal generated, the next stage is first to move it to the required frequency and then amplify it to the required level. Conversion to the right frequency may be undertaken in two or even three stages as it is in many receivers. One of the mix processes will involve the use of a variable-frequency oscillator. Today this is most likely to be a synthesised oscillator, the frequency and operation of which is under microprocessor control. This gives the opportunity to have facilities like dual VFOs and the like, even though only one synthesiser is used. After each mix stage, careful filtering is required to ensure that the level of spurious signals is kept acceptably low. Once the signal is on the right frequency it is amplified to bring it to the required power level. For most transceivers today this is around 100W output.

The QRP transmitter required a filtering and matching network at the output of the power amplifier, and the same is true for a higher-power Morse and single sideband transmitter. In fact, the level of spurious signals is even more important because power levels are generally much higher and the possibility of causing interference is that much greater.

VOX and break-in

A system known as *press-to-talk* is normally used to change a transmitter from receive to transmit. This is normally accomplished with a *pressel* (press switch) on the microphone. (When using Morse a transmit/receive switch can be employed.) However, under many circumstances it is convenient to be able to speak and let the electronics detect the presence of sounds, automatically switching from receive to transmit. This system, known as *VOX*, is included on many sets.

While the system may appear to be very convenient, it is not quite as foolproof as it may appear. First, any noise of a sufficient level may trigger the system and this can result in the transmitter changing over at the wrong time. There is normally a control to set the level at which the changeover should take place but even this does not give complete immunity from unwanted changeovers. Second, even sounds from the receiver loudspeaker can find their way into the microphone and cause the transmitter to transmit. This can be overcome with a control that is normally labelled 'Anti-vox'. This effectively nulls out the receiver sounds so that they do not affect the vox operation. Finally, it takes a finite time for the receive/transmit changeover to take place. If relays are used in the changeover the time taken may result in the first syllable being clipped. This can often be detected at the receiving end, and can sound a little odd. As a result most DX operators prefer to use press-to-talk and have the operation completely under their control, as unwanted or spurious changes can result in them missing receiving important information.

For Morse operation the situation is a little different. Virtually all key presses are likely to be intentional and as a result *break-in*, as it is called, is far more useful. However, it is very important that changeover is fast, otherwise the first dot or dash will be shortened. Under some circumstances, when high speeds are used and the changeover is slow, it can result in the first dot being missed. Some transceivers use all-electronic switching in the changeover and can accomplish the switch from transmit to receive and back again very quickly. In these instances the response of the receiver AGC to the changes is the main factor. When this is well designed, full break-in is very useful, offering the ability to transmit by simply pressing the key. It also enables listening between the dots and dashes, giving a much greater 'visibility' of what is happening while you are transmitting. This can be an enormous advantage when many people, including you, may be calling a DX or contest station at the same time.

Keying

Most new transceivers contain keyer electronics and simply require a paddle to be connected to provide a fully *iambic-mode keyer* (ie a fully automatic keyer that generates a sequence of interspersed dots and dashes when both paddles are squeezed together). While connections are provided for normal keys and keyers as well, the fact that keyer electronics are included in the set means that the cost of buying an external keyer is avoided and only the paddle is required.

Speech processing

It has been said that single sideband without the use of speech processing is an outmoded form of communication. The reason for this is that the output from a single sideband transmitter is limited by the peak audio level that is being transmitted. If the average power is very low, then the signal will be perceived as being relatively weak. If the average power of the signal can be raised then the signal will appear to be much stronger and will be able to be heard more easily through interference and above the noise. Unfortunately speech has a very high content of short-lived peaks and a low average power level. To make the best use of the transmitter power the average level of the

HF AMATEUR RADIO

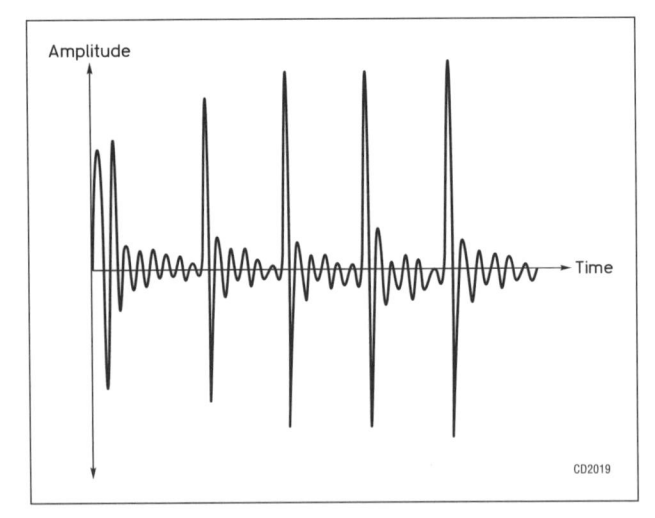

Fig 5.4. A typical speech waveform, showing that human speech has a very low average power. As the transmitter is limited by the peak level it can carry the average level of the modulation must be increased to use the available power capability more efficiently

audio has to be raised using a speech processor.

There are a number of reasons why speech has a low average power. First, the intensity of the speech may vary – the speaker may place more emphasis on one word than another or may move away from the microphone. In this way the overall level of the speech varies. This can be very significant and even when the speaker is trying to maintain a constant level it can vary by 10dB. Another reason for the low average level is that speech contains short-lived peaks or transients. Sounds like 'p' and 'b' are known as *plosive* sounds and start with a burst of energy after which the level falls away. Other sounds too, have large peaks, whereas the vowel sounds tend to be more constant.

In addition to this, not all frequencies contribute as much towards the intelligibility of a signal. The best use can be made of the available power by ensuring that only the frequencies that contribute most to the intelligibility of the signal are included.

Processing is achieved in a number of ways: compression, clipping and frequency tailoring all play their part. Some of the more sophisticated processors use all three methods and can give up to 8dB gain – more than most linear amplifiers can supply and at much less cost.

Speech *compression* is a form of processing where the gain of an amplifier is steadily reduced as the audio level increases. Generally the gain reductions only occur once a certain level is reached. The gain may be reduced instantaneously so that it varies over the audio waveform itself.

Another form of compression has a time constant introduced so that it acts like an audio AGC and alters the gain according to the peak level of the signal (Fig 5.5). In this way the general level of the signal is maintained at a given level. This type of circuit is often called *VOGAD* (voice-operated, gain-adjusting device). The time constants in the circuit are important. The

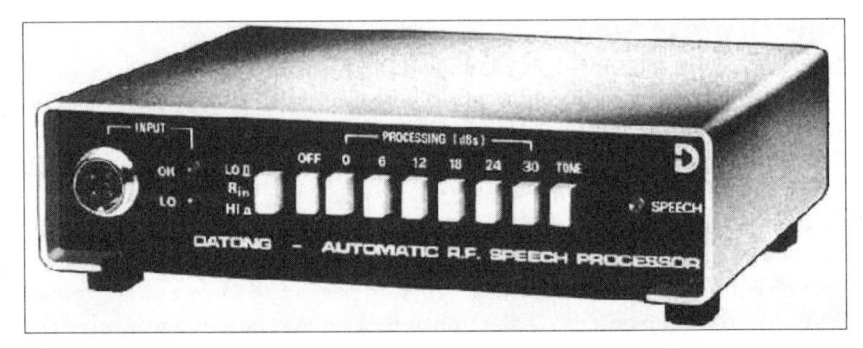

A typical example of an external speech processor

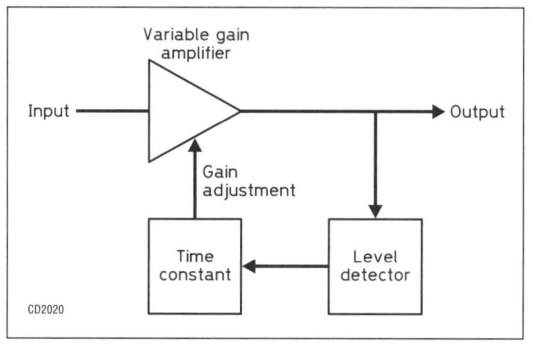

Fig 5.5. Block diagram of a compressor using a time constant. This type of device is often called *VOGAD* (voice-operated, gain-adjusting device) and it is used to maintain the general level of the signal

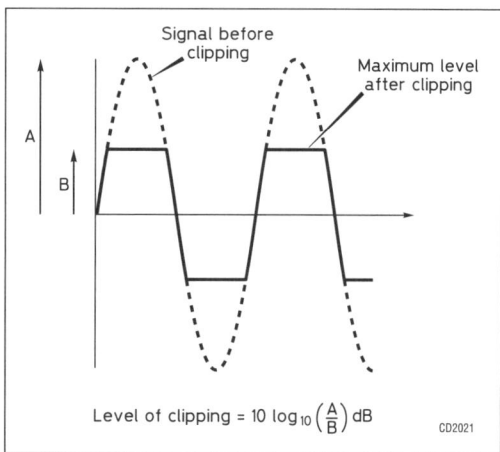

Level of clipping = $10 \log_{10}\left(\frac{A}{B}\right)$ dB

Fig 5.6. The level of clipping is the ratio between the peak levels before and after clipping expressed in decibels

attack time should be fast to ensure that any sudden increases in level do not pass beyond the compressor and cause any overloading – typically speeds of around 10ms are used for this. The decay time should be much slower so that the gain level follows the level of the speech – usually decay times of around 300ms are employed.

Clipping may also be employed (Fig 5.6). This technique involves removing the peaks of the waveforms. In this way the peak-to-average ratio can be considerably improved. While it may appear that removing part of the waveform might remove vital information from the waveform and reduce the level of intelligibility, this is not the case if the clipping is performed well. The reason for this is that the sounds are recognised primarily by their frequency content.

The level of clipping is important when clipping a signal. A figure is often quoted in decibels and this refers to the ratio between the peak levels of the signal before and after clipping.

Unfortunately the clipping process is not linear by its very nature, and distortion products are introduced into the signal, most of these being in the form of harmonic distortion (Fig 5.7). Typically signals for communications purposes are limited to 3kHz, and these distortion products will cause the frequency content from the clipper to extend beyond this limit. This means that a clipper of this type would need to have a filter after it to remove these products. Unfortunately many frequencies in the wanted signal give rise to harmonics that fall within the wanted audio bandwidth, and these have the effect of reducing the level of intelligibility. As a result the level of clipping that can be employed for 'baseband' (audio) clipping is limited to about 10 to 15dB – beyond this

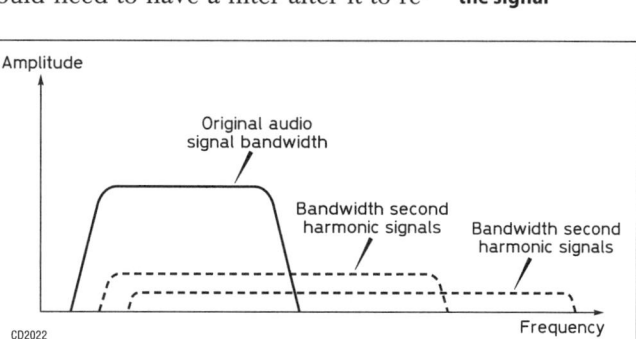

Fig 5.7. Harmonic distortion products created during audio clipping. Those above the maximum required frequency can be removed by filtering, but those in the required audio bandwidth cannot be removed and reduce the intelligibility of the signal

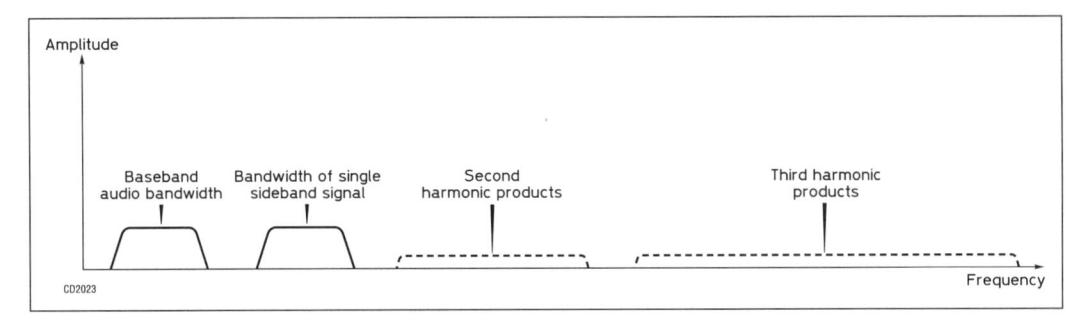

[Figure: spectrum showing Baseband audio bandwidth, Bandwidth of single sideband signal, Second harmonic products, and Third harmonic products plotted against Frequency.]

Fig 5.8. By clipping a radio frequency signal the harmonics caused by clipping fall at multiples of this signal and can be filtered out very easily, thereby reducing the level of distortion and allowing greater levels of clipping to be achieved

the level of distortion that is produced starts to reduce the intelligibility of the signal. This limits the actual gain that can be achieved by an audio processor to around 5dB.

To overcome the problem a radio frequency signal can be used because when this is clipped the harmonics occur at multiples of this signal and can be easily filtered out (Fig 5.8). To achieve this a single sideband signal must be generated. A signal with good sideband and carrier suppression is required to reduce intermodulation products that would impair the overall performance. The signal is clipped at the radio frequency, filtered to remove the out-of-band products including harmonics and then demodulated to regenerate the clipped audio. In some transmitters the clipper is included in the sideband generator to provide an integrated processor within the transmitter. In this case it is obviously not necessary to include the demodulator stage of the processor.

While RF clippers are obviously more complicated than their AF counterparts they can produce significantly more gain. As already mentioned, the maximum gain that can be usefully obtained with a AF processor is around 4 to 5dB, whereas an RF processor can give up to about 8dB. Every decibel of gain that can be obtained is useful in helping make the station more effective and make contacts through pile-ups or under low-signal conditions.

Useful improvements can also be made by ensuring that the frequency content of a signal is tailored. Even landline telephone systems have their frequency response limited to cover frequencies between 300 and 3000Hz. Communications systems normally tailor this even more, limiting the top frequency to 2.7kHz or even slightly less. While this does reduce the natural sound of the signal, there is only a marginal drop in intelligibility.

Reducing the low-frequency content of the signal and pre-emphasising the higher frequencies is found to reduce an effect that occurs in clipping where the stronger low-frequency components are emphasised, making the signal sound rather unnatural. Typically a response that falls below 1kHz may be useful.

Many processors today combine all three methods (Fig 5.9). On entry to the processor the signal may have its frequency bandwidth limited and the low-frequency content reduced. The signal may then enter a VOGAD stage to ensure a constant audio level is presented to the clipper. After clipping, preferably at RF, the signal is filtered and in the case of an RF filter demodulated before being presented to the transmitter audio input.

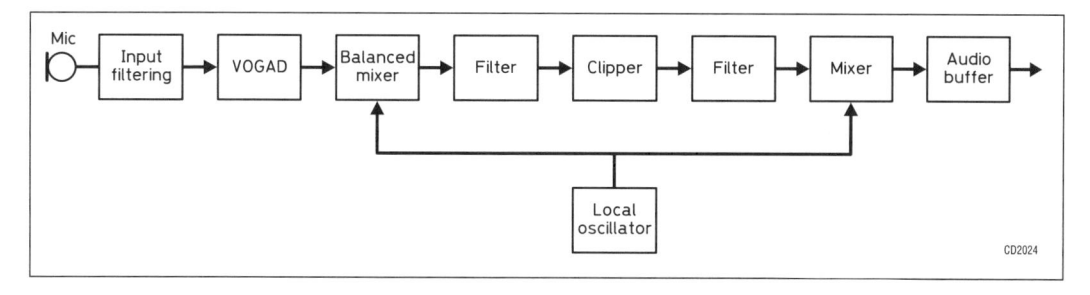

Fig 5.9. A speech processor combining several forms of processing to give the optimum performance

Processors are particularly useful because they enable the best use to be made of the available RF. In some cases they may be used instead of a linear, providing a very cost-effective solution.

Linear amplifiers

Usually HF transmitters are capable of providing around 100W output power. In most countries the maximum legal limit is somewhat higher and to reach it a further amplifier is needed. Amplifiers used for single sideband must be linear – if they are not, the waveform becomes distorted and the signal will splatter over the band, causing interference to other users.

As amplifiers are often manufactured so they can be used anywhere in the world, it is possible that they can produce power in excess of the legal limit for the country where they are being used. Accordingly, care should be taken to ensure this limit is not exceeded.

While linear amplifiers can be expensive, they can provide a useful amount of gain. In the UK an output power of 400W (26dBW) may be used by Class A licensees (information correct at the time of going to press). When compared to a transmitter with an output of 100W, the station using the

A typical example of a linear amplifier

linear will be 6dB higher in level. This is a significant amount and could make the difference between making a contact or not. For example, in a pile-up of competing stations it can make all the difference between you making contact with the DX station or being left calling with all the other weaker stations. While a linear is not an essential piece of equipment, it can certainly make a difference, especially when the going gets tough.

Transceivers

The greatest number of units which are sold these days are not individual receivers or transmitters, but transceivers containing all the circuitry for transmitting and receiving. Obviously a single unit is far more convenient than having separate units. Also the cost of a transceiver is less than the sum of its individual counterparts because a large portion of the circuitry can be used for both transmit and receive functions. Stages like the mixers, local oscillator, IF amplifier and so forth can all be used in both functions.

In order to make a transceiver, the routing of the signal has to be changed between transmit and receive. This can be done using either relays or semi-conductor switches such as PIN diodes.

Transmitter power

There are a number of methods of making power measurements. Some years ago it was common to measure the DC power input to the final amplifier. This method was adopted because it was easy to measure the voltage supplied to the final amplifier and the current drawn. It was also easy because the current was steady for Morse and the AM transmissions that were being used at the time.

These days power is normally measured at the output of the transmitter. While many measurements are made directly in watts it is more common to define power in terms of 'dBW', and indeed the UK licence quotes power levels in these terms. Although dBW may not be quite as easy to use, its concept is quite straightforward. It is simply the power expressed in decibels relative to one watt. To give an example, a power of 10W is 10dB increase on 1W, and so it can be described as a level of 10dBW. Similarly a power of 400W is 26dB above 1W and this means it can be expressed as 26dBW.

While it is relatively easy to measure the power of a steady waveform such as that of a Morse signal with the key held down, the same is not true of a single sideband signal. The output power varies in line with the instantaneous level of the audio. To accommodate this, the power of a single sideband signal is measured at the peak of the radio frequency signal envelope. Accordingly this power is called the *peak envelope power* (PEP).

Spurious signals

With many HF transmitting stations radiating significant levels of power it is necessary to ensure that they do not generate any undue levels of spurious signals that might cause interference to other users. There are a number of causes of spurious signals – one is when harmonics are generated, another arises from intermodulation distortion, and a third arises from unwanted mix products being insufficiently filtered and thus being present at the output.

Intermodulation

One of the major causes of this type of interference arises out of the poor linearity of amplifier stages in the transmitter, and in particular the final amplifier. It gives rise to intermodulation products in the same way that it does in a receiver front-end amplifier. The products arise when two or more signals are passed through the amplifier. While this distortion will not occur in the case of a Morse signal where only one carrier is present, a single sideband signal consists of a whole variety of different frequencies within the transmitted bandwidth.

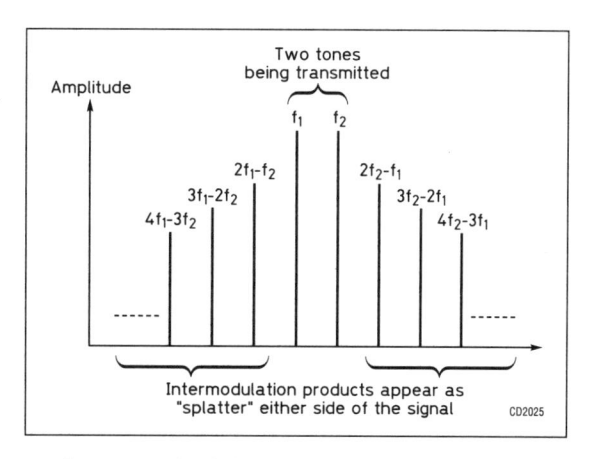

Fig 5.10. Intermodulation products generated in a linear amplifier

When specifications are given they take the case where two tones or closely spaced signals are passed through the amplifier. To illustrate the way in which this is done, imagine a single sideband signal which is derived from just two audio tones, one at 1kHz and the other at 2kHz. The resulting sideband signal will be two radio frequency signals spaced by 1kHz. When passed through an amplifier these will give rise to third-order and higher-order intermodulation signals in the same way that was described for a receiver amplifier in Chapter 4. In the case of the receiver we saw how the third-order mix products gave signals at $2f_1 - f_2$, $3f_1 + 2f_2$ etc and $2f_2 - f_1$, $3f_2 + 2f_1$ etc as shown in Fig 5.10.

In the case of a real single sideband signal, there will be a whole variety of different audio frequencies. All these intermodulate with one another to generate noise or splatter which spreads out from the main signal. Normally the worst intermodulation products will be those which are nearest to the wanted signal, and their levels reduce as the offset increases.

The specifications for intermodulation products are usually given in terms of the difference (in decibels) between the wanted or main signal and the various intermodulation products. However, a transmitter specification will often say that all intermodulation products are below a given level, and in this case the worst ones are bound to be the third-order products. Sometimes (especially in a review) the levels of specific products will be stated.

Typically values are around −25 to −30dB for the third-order products and five or six decibels lower in the case of the fifth-order products (the more negative the number, the better the performance). A typical modern transmitter should have all its products better than −25dB relative to the main signal.

Harmonics and other spurious signals

Harmonics of the required frequency are also generated in a transmitter. Other signals may also appear at the output. These may arise from the mixing process in the transmitter. Whatever the cause they need to be kept to acceptable levels to ensure that no undue interference is caused to other users.

The level of a spurious signal is not normally given in absolute terms, ie a given number of watts or milliwatts. The more usual way is to relate it to the level of the wanted signal. In other words, a spurious signal will be said to be a certain number of decibels below the carrier, sometimes measured in 'dBc' units. Again, the more negative the number is, the smaller the spurious signal level. For example, a spurious signal level of −40dBc is better or lower in level than one of −31dBc.

The ideal level for spurious signals would be zero but this is obviously impossible. Normally transmitters are specified as having all spurious signals at a level which is at least 40dB below the carrier, ie −40dBc. Often the specification is left rather vague because the levels of harmonics and other spurious signals will vary according to the band or frequency in use. Typically the levels of harmonics may be at levels of −50dBc or even as much as −60dBc.

In cases where interference is being experienced on frequencies above 30MHz it may be necessary to add a low-pass filter to an HF transmitter. These filters are available from many amateur radio stockists and give very high levels of out-of-band attenuation. If fitted correctly they will reduce any above-band harmonics to well within acceptable limits.

VSWR tolerance

When transmitters operate under conditions where there is a high level of standing waves, high voltages may be present at the power amplifier, or it may be required to provide very high levels of current. These could cause the output devices to operate outside their ratings and may result in their destruction. To prevent this happening most transmitter power amplifiers incorporate protection circuitry. When a high level of VSWR is detected, it causes the output from the transmitter to be reduced so that the safe operating conditions for the output power devices are not exceeded. Generally, a VSWR level of 2:1 is acceptable but 3:1 may cause a power reduction to be noticed.

Incremental tuning and separate VFOs

Most transceivers today have very flexible tuning facilities. The exact way in which these may operate depends on the particular transceiver in question and may include *incremental receiver tuning* (IRT) and multiple VFOs.

Incremental receiver tuning is normally used to give a small amount of receiver shift while retaining the transmitter on the same frequency. This is very useful when needing to 'clarify' the incoming signal by changing the receiver tuning very slightly. If the main tuning control were used then it would also change the transmitter frequency as well. The receiver incremental tuning control is generally an additional knob that normally gives a shift of a few kilohertz at the most.

The majority of modern transceivers have a twin VFO facility. This enables the tuning to be set to one channel on one VFO. This can be left on a channel that may be required later and the other one switched in and used. Switching between the two VFOs is normally very flexible, and almost invariably it is possible to receive on one VFO and transmit on the other. In this way split-frequency operation can be undertaken. This is particularly useful because many DX stations transmit on one frequency and receive on another a few kilohertz off-channel to ease operation and manage a pile-up when many stations are calling. Typically the transmit and receive frequencies may be between 1 and 5kHz apart when using Morse and between 5 and 20kHz apart when using sideband. This degree of shift is often beyond the scope of the incremental receiver tune facility and the use of twin VFOs is far easier and more flexible.

Automatic level control

As the intermodulation performance depends largely upon the linearity of the final amplifier, it is imperative that the latter is not over-run. When this happens it will start to limit and there will be a marked increase in the distortion and hence the amount of 'splatter' being picked up by nearby receivers. To prevent this, transmitters use *automatic level control* (ALC) which detects the level of the signal at the output and reduces the gain of previous stages to prevent overload. In fact it operates in the same way as AGC in a receiver. The main point to note is that no transmitter amplifier should be run close to its limits, otherwise distortion levels rise and cause interference to other users.

CHAPTER 6

Antennas

In this chapter:

- Feeders
- Standing wave ratios
- Antenna tuning units
- Directivity
- End-fed wire
- Dipole
- Multiband antennas
- Vertical
- Beam (Yagi)
- Earth connections
- Selecting a location
- Safety

THE performance of any antenna is very important and will greatly affect that of the station, so time and energy invested in it will pay dividends when operating on the bands. A poor antenna will limit the performance of the station, however good the transmitter and receiver or transceiver may be. Conversely a good one will make the most of the equipment to which it is connected.

Unfortunately it is not usually possible for us to have the antenna system we would like. The garden may be too small or there could be other restrictions that prevent it being installed. While there is no doubt that stations with better antennas are able to contact more DX stations, this does not mean to say that all is lost if a large system cannot be deployed. Often disadvantages in one area can be made up in others, and many people enjoy experimenting with antennas to find the best one for their particular location.

Constituents of an antenna system

An antenna system is made up of a number of parts, not just the antenna itself. A feeder is required to enable the power to be transferred to the antenna from the station or vice versa. There may also be some matching circuitry between the feeder and the antenna – this is required because the maximum amount of power is transferred from the feeder to the antenna when the impedance of the two is the same.

Feeder

Feeders (transmission lines) are used to transfer RF energy from one point to another, and they should introduce as little loss as possible. They transfer the signals by propagating a radio frequency wave along their length while not allowing it to radiate. This is done by confining the electric and magnetic fields associated with the wave to the vicinity of the feeder.

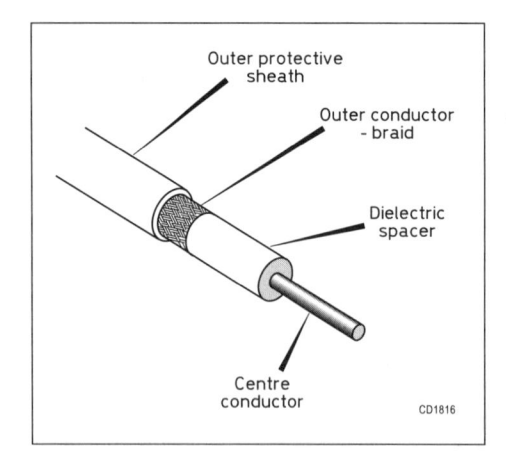

Fig 6.1. Coaxial feeder

There are several forms of feeder. The type that is the most common is *coaxial feeder* or 'coax' for short. It consists of an inner conductor surrounded by an insulating dielectric and covered with an outer screen or braid, which in turn is covered with insulation to act as protection. It carries current in both inner and outer conductors, but because they are equal and opposite all the fields are confined within the cable and cannot radiate. As there are no fields outside the cable, nearby objects do not affect its properties and it can be used to carry RF energy through many locations with little risk of them being affected.

Another type of feeder that is sometimes used on frequencies below 30MHz consists of two parallel wires. As the currents that flow in the two wires is equal and opposite, in theory no signal should be radiated from them. In practice it is found that they are affected by nearby objects and this form of feeder should not be run through a house for example.

There are a number of types of twin feeder. The most common consists of two wires spaced about 20mm apart covered in a translucent plastic and is called *twin feeder* or *ribbon feeder*. Whilst it is fine for internal use, when used externally the plastic dielectric can absorb water which causes losses to rise significantly. An alternative version using black plastic with ovals cut in the spacing dielectric is intended for external use and works well.

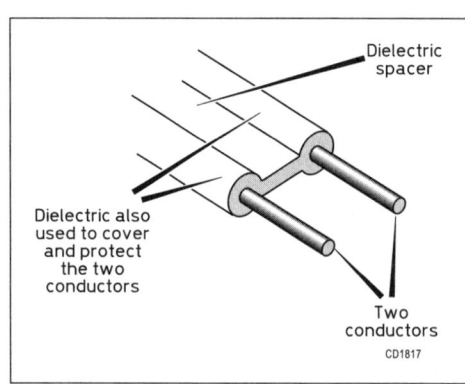

Fig 6.2. Twin or ribbon feeder

It is possible to make *open-wire feeder*. This can be achieved by taking two lengths of wire, separated at intervals by spacers. These can be made from plastic waste piping (the white variety), which is cut into lengths of about 15cm. This is then cut into strips with holes drilled at either end and attached to the wire as shown in Fig 6.3. The spacers can be placed every metre or so, dependent upon how well the wires need to be spaced at that particular point. Obviously the more spacers that are used, the better the control of the spacing. This type of open-wire feeder can be made very

CHAPTER 6: ANTENNAS

easily and when used away from other objects gives very low levels of loss. Its impedance will typically be very roughly around 600 ohms and this is often quite suitable for feeding an antenna when combined with an antenna tuning unit. Any imperfections in the match can easily be tuned out.

The loss that a feeder introduces between the antenna and receiver or transmitter is of the utmost importance, as it will reduce the efficiency of the station. This loss occurs in two main ways. The first is as power radiated from the feeder which, although normally small, can reach measurable proportions especially over long runs. The second is as power being dissipated as heat as a result of the resistance of the conductors of the feeders. These ohmic losses can be reduced by making the conductors larger so that their resistance is lower.

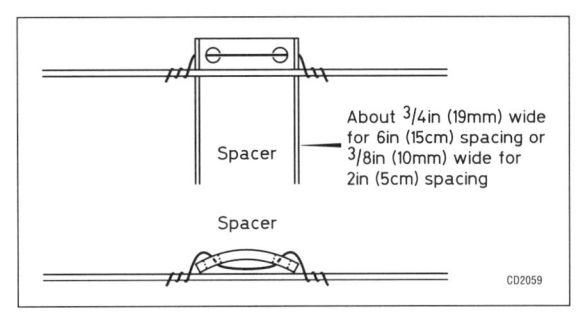

Fig 6.3. An example of simple and inexpensive construction of 600-ohm open-wire feeder. Plastic piping of 5 to 8cm diameter is cut into short lengths, and then sawn lengthwise into strips, a pair of holes being drilled near the end of each strip. Spacing may be 5 to 15cm with insulation at intervals of some 12 to 15 times the line spacing. Losses are reduced with wide spacing and construction is much quicker, but symmetry is more easily upset. Note that the spacers are slightly curved and this helps to prevent slippage

The dielectric between the conductors can also give rise to losses. Power may be dissipated here, especially if moisture is present. It is therefore important to ensure that moisture does not enter coaxial cable.

Characteristic impedance and velocity factor

It has already been mentioned that the maximum power transfer from a transmitter to an antenna occurs when their impedances are both the same. Feeders also have an impedance and this is of great importance when designing and erecting antennas.

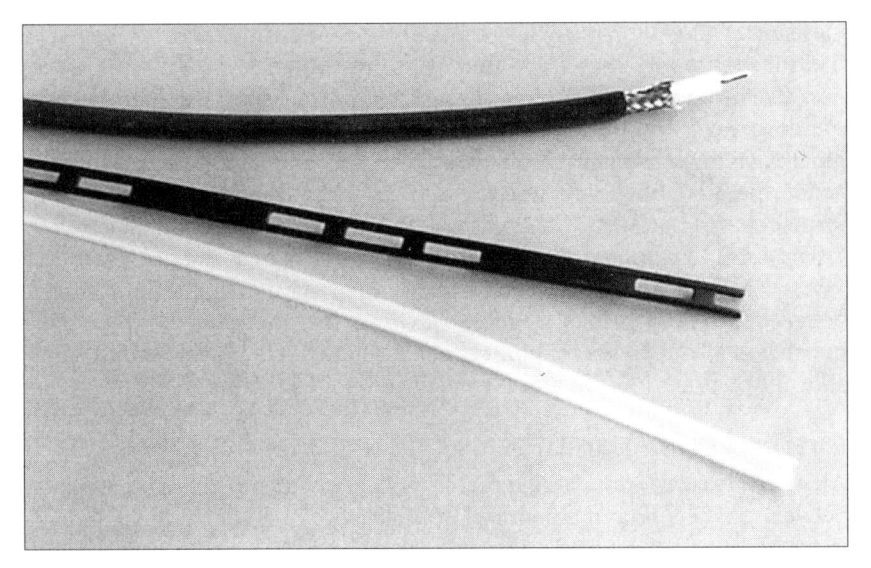

Coaxial cable, black twin and ordinary twin feeder

65

HF AMATEUR RADIO

The impedance of a feeder can be best demonstrated by taking the example of an infinitely long line with no losses. A signal applied to this line would propagate along it forever and never be reflected or returned. As the energy propagating along the wire is always travelling and not stored the line would look like a pure resistor to the transmitter.

If the line was cut at a finite distance from the transmitter and the end left either open-circuit or short-circuited then any signal travelling along the feeder would not be able to travel any further. In this case the only way for it to travel is to be reflected back the way it came.

Now if a variable resistor with its value set higher than the characteristic impedance of the line is connected to the end of the line, some power will be dissipated in this resistor. As the value of the resistor is reduced it is found that more power is transferred to it and less is reflected. Eventually a point is reached where all the power is dissipated in the resistor and none is reflected. If the value of the resistor is reduced beyond this point less power is transferred to it and more becomes reflected.

The value of the resistor when no power is reflected represents the impedance of the feeder itself and it is known as the *characteristic impedance*. Generally coaxial cables have an impedance of 75 ohms if they are to be used for domestic television, or 50 ohms if they are for professional, amateur radio or CB use. Other impedances are often used for computer applications so beware if any of this type of cable is used.

Standing wave ratio

When a load (or in this case an antenna) is not properly matched to the feeder then not all the power reaching the end of the latter can be transferred to the load. However, the power cannot just disappear and has to go somewhere. The only place that it can go is back along the feeder, which gives rise to a number of effects. High voltage and current points are set up along the feeder and, especially when high-power transmitters are used, these may cause damage to it. These points may also appear at the output of the transmitter and may cause damage to the output devices if no protection circuitry is present. If this circuitry is present, then the high level of standing waves will be detected and the transmitter power may be reduced. Finally, the high level of standing waves may mean that the antenna system is not operating efficiently and it can be optimised to give improved performance.

The level of standing waves is normally quoted as a *standing wave ratio* (SWR). A ratio of 1:1 shows that there is a perfect match and no power is being reflected. A ratio of infinity to 1 shows that there is a complete mismatch and all the power is being reflected. A ratio of 2:1 is normally acceptable although it is best to aim for the minimum attainable.

Antenna tuning unit

An *antenna tuning (matching) unit* (ATU) is used in order to give a good match between the feeder and antenna. These are widely available, consisting of variable capacitors and inductors, and are an essential part of many antenna systems. Ideally the ATU is placed at the interface between the feeder and the antenna itself, but this is often not practical. In some antenna systems, such as the open-wire-fed doublet or the end-fed wire, the ATU may be placed between the coaxial feed from the transmitter and the end of the open-wire feeder or wire antenna. Where coaxial feeder is used all the way from the transmitter to the antenna, an ATU located close to the transmitter can be used to reduce the SWR seen by the transmitter's power amplifier. In this way it may enable the transmitter to reach its full output as many sets include a system that will reduce the power output under conditions of high SWR to protect the output devices.

Directivity

An important aspect of any antenna is the way in which it radiates in different directions. No antenna radiates equally in all directions, and as a result the radiation pattern around it is of interest – a plot of this pattern is known as a *polar diagram*. When talking in terms of antenna patterns it is often easiest to visualise the way in which they are generated in terms of a signal being transmitted, but an antenna will act in exactly the same way when receiving a signal.

Diagrams for two types of antenna are shown in Fig 6.4. The first example is that of a half-wave dipole. This has a figure-of-eight shape, radiating most of the signal at right-angles to the axis of the antenna and least along its length. Other antennas have more pronounced radiation patterns. Some types of antenna are designed specifically to 'beam' power in one particular direction and are therefore called *beams*. The same is true when receiving and in this case the antenna is more sensitive in that direction, making them ideal for DXing where this effect can considerably help reception.

As more power is radiated in some directions the antenna is said to have *gain* in the direction of its maximum radiation. It should be remembered that antennas with gain will radiate less in other directions, which can be used to advantage as it can ensure that the transmitted signal is concentrated only in the direction where it is needed, reducing the level of interference that may be caused to stations in other directions. Conversely, when receiving the level of interference caused by stations that are not on the direction of the beam is reduced. This can be a considerable advantage when the band is crowded with strong stations.

The gain of the antenna is expressed in decibels and is usually compared to a dipole antenna where the gain is expressed as dBd (decibels relative to a dipole). In some instances the gain will be compared to an *isotropic source* (a theoretical antenna that radiates equally in all directions) and expressed as dBi. Beware when comparing gain for different antennas because figures

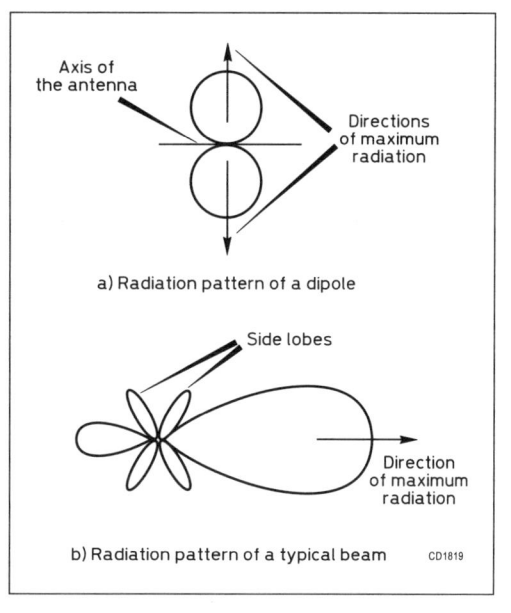

Fig 6.4. Radiation patterns of two types of antenna

compared to an isotropic source are 2.1dB higher than those compared to a dipole and either reference may be used in a specification.

Often equally important as the forward gain is the *front-to-back ratio*. This figure is the ratio between the power in the forward direction of the antenna to the power in the reverse direction and this is expressed in decibels. When making adjustments to an antenna, it is found that the optimum forward gain and front-to-back ratio do not usually coincide. In many instances people will choose the maximum forward gain but, when high levels of interference are normally experienced in the direction opposite to the wanted stations, the maximum front-to-back ratio may be the best option.

It should be remembered that when a directional antenna is used, it should have some means of orientating it in the required direction. Most stations using a beam will use a *rotator*, which is a unit that has a motor on the masthead to physically rotate the beam, and a controller that can be located by the equipment. This controller remotely controls the direction of the antenna, allowing it to be orientated in the required direction. The rotator and the associated mast mean that the total cost of a rotatable antenna for the HF bands is considerably more expensive than the cost of the antenna alone, and this factor should be remembered when considering such an antenna system.

End-fed wire

One of the easiest forms of antenna to erect is an *end-fed wire*. It simply consists of a wire from the transmitter and receiver strung between two points which are as high as possible. Often it may be of random length but in many cases it can be cut to a quarter-wavelength or a multiple of quarter-wavelengths. This will mean that the antenna is fed at a current maximum point that is much easier to match than one where there is a voltage maximum.

A wire of this nature is very easy to install, requiring little in the way of special feeders. If it is only to be used for receiving then it is possible to connect it just to the input of the receiver, although for the best performance an antenna tuning unit is required. This will also be essential if the antenna is to be used for transmitting.

A typical installation is shown in Fig 6.5, although the exact nature of the antenna will depend very much upon the individual situation. In this instance the wire leaves the antenna tuning unit and passes out of the house. It may be possible to drill a neat hole in the window frame, making sure to seal the outside to prevent moisture entering the latter.

With the installation shown the vertical section travels up to meet the horizontal section. Nylon rope can be attached to either end, and insulators used. These can be bought from amateur radio stockists, and provide a neat way of terminating the wires as shown. Proper fixings should be used to attach the rope to the house so that it does not wear against the brickwork.

It is often convenient to use a tree as one of the anchor points for the antenna. If this is done then provision must be made to accommodate any movement due to the wind. This can be achieved by using a system like that

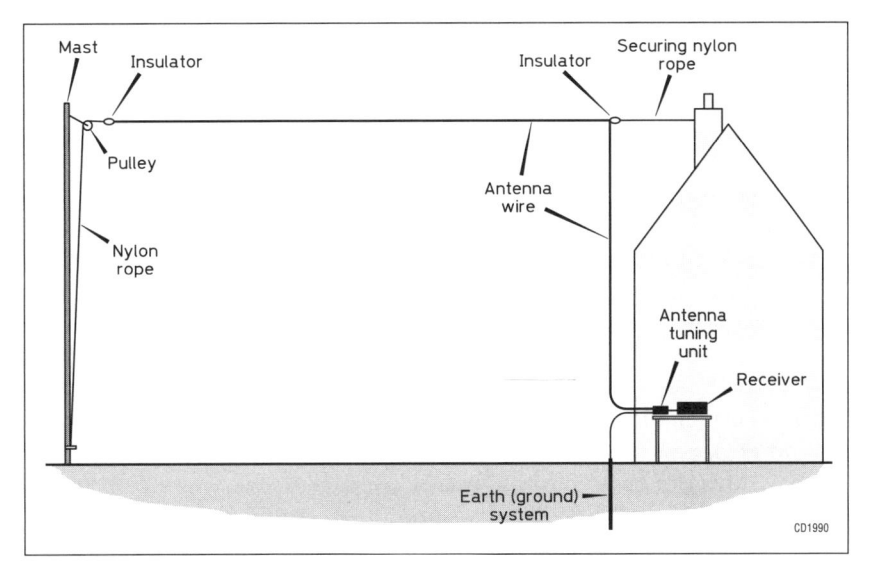

Fig 6.5. A typical end-fed wire installation

shown in Fig 6.6. Here a constant tension is held on the wire regardless of the movement of the tree. However, there should be sufficient rope in the loop to accommodate the movement under the windiest conditions, and care should be taken to ensure that the loop will not become snagged in the tree, thereby rendering it ineffective.

It must be said that this type of antenna is far from ideal. It radiates along its whole length and this means that there are likely to be high levels of radio frequency radiation in the shack that could be a hazard, particularly if high power levels are used. It can also cause interference to household electronics or to other pieces of equipment in the shack – sometimes RF can be picked up by the microphone input to the transmitter, causing the audio to become distorted.

For an end-fed wire to operate satisfactorily it should have a good earth system. In many instances this is given less attention and the effectiveness of the whole antenna is reduced.

Often the name 'long wire' is used for these antennas. Although the term normally applies to an ordinary end-fed wire, it actually refers to an antenna that is several wavelengths long. It is found that an antenna like this is highly directional, being what is called an *end-fire antenna*. As the name suggests, the direction of maximum radiation is along the axis of the antenna itself.

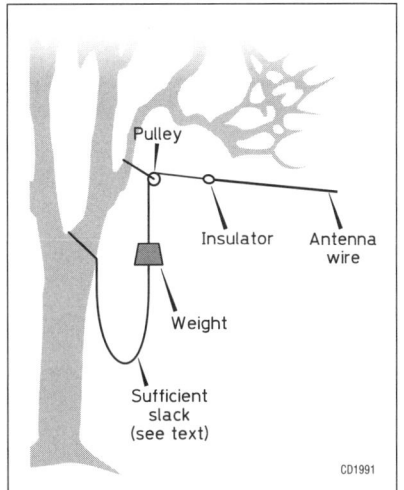

Fig 6.6. Anchoring a wire antenna to a tree

Dipole

The dipole is one of the most important types of antenna. It is widely used in its own right, and forms the basis of a number of other antennas that are very popular.

In its most common form the dipole is an electrical half-wavelength long, and fed in the middle as shown in Fig 6.7. For completeness it should be mentioned that although it is usually thought of as a half-wave antenna it can be made any number of half-wavelengths. This means that an antenna that is cut to operate as a half-wavelength dipole on one band will also operate as a three-half-wavelength dipole on a band that is at three times the frequency of the lower one. For example, a half-wavelength antenna for 7MHz can be used as a three-half-wavelength antenna on 21MHz.

The polar diagram of an antenna varies according its length in wavelengths. A half-wave dipole has the typical figure-of-eight polar diagram but as the length is increased it develops lobes which tend to align progressively with the axis of the antenna as its length is increased. This can be seen from Fig 6.8.

The impedance of the antenna is important. A dipole operating in free space has a feed impedance of 73 ohms but nearby objects, including the ground, will alter this, and 50-ohm cable will usually give a good match. Also, when it is used with other elements in an antenna the impedance drops and special matching arrangements may be needed.

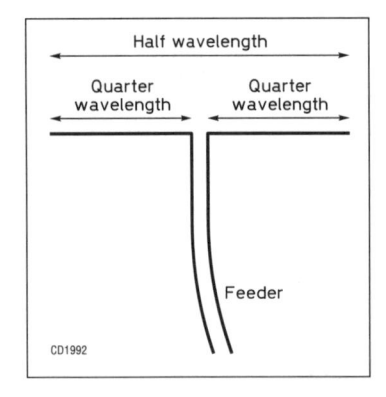

Fig 6.7. A half-wave dipole

The length of the antenna is quite critical if it is to work properly – it must be an electrical half-wavelength (or multiple of half-wavelengths). This length is not quite the same as the wavelength in free space. There are a number of reasons for this; one is called the *end effect* and is due to the fact that the antenna is not infinite. The length of a half-wave dipole can be calculated from the formula:

$$\text{Length(m)} = \frac{148}{\text{Frequency(MHz)}}$$

$$\text{Length(ft)} = \frac{480}{\text{Frequency(MHz)}}$$

Using the formula it can be seen that lengths for antennas in the HF bands are as shown in Table 6.1. However, it should be remembered that these are approximate lengths and the antenna should be cut slightly longer first and trimmed to give the optimum operation in the section of the band where it is required.

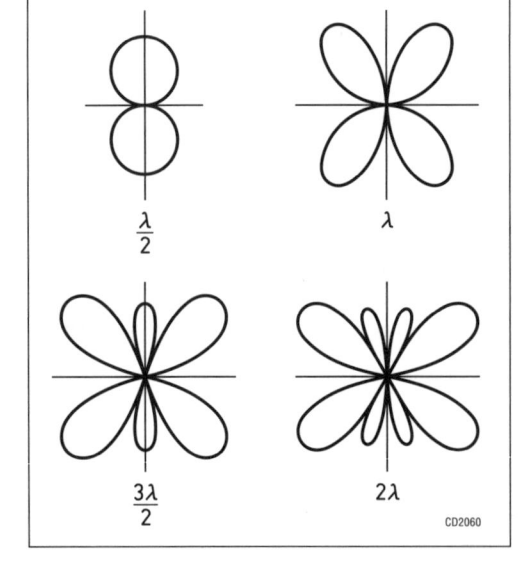

Fig 6.8. Polar diagrams for half-wavelength and three-half-wavelength antennas

Building a dipole

The most straightforward way to install a dipole is as a horizontal antenna, although this is by no means the only way and various other methods and types will be described later. The basic dipole is shown in Fig 6.9. As with the end-fed wire, some means of strain relief is required to account for any

CHAPTER 6: ANTENNAS

movement if a tree is used as one of the anchor points. Again, insulators should be used at either end and nylon rope can be used to secure the dipole itself.

The centre of the dipole requires the coaxial or open-wire feeder to be connected to it. While it may be tempting to simply connect the feeder and let it take the strain, this is not particularly satisfactory when there is a long drop for the feeder – a dipole centre should be used. This will take the strain caused by the tension on the wire, thereby avoiding damage to the feeder over a period of time. Often it is possible to use an ordinary antenna insulator for this purpose.

While it may not always be possible it is certainly helpful to route the feeder away from the antenna at right-angles to the antenna wire as shown. In this way it will have the minimum effect on the operation of the antenna.

Often a *balun* is placed at the feed point of the dipole. This is a transformer used to connect a balanced system to an unbalanced one, or vice versa. It is required because a dipole is a *balanced* antenna, ie neither connection is earthed, and coaxial feeder is *unbalanced*, having the outer braid

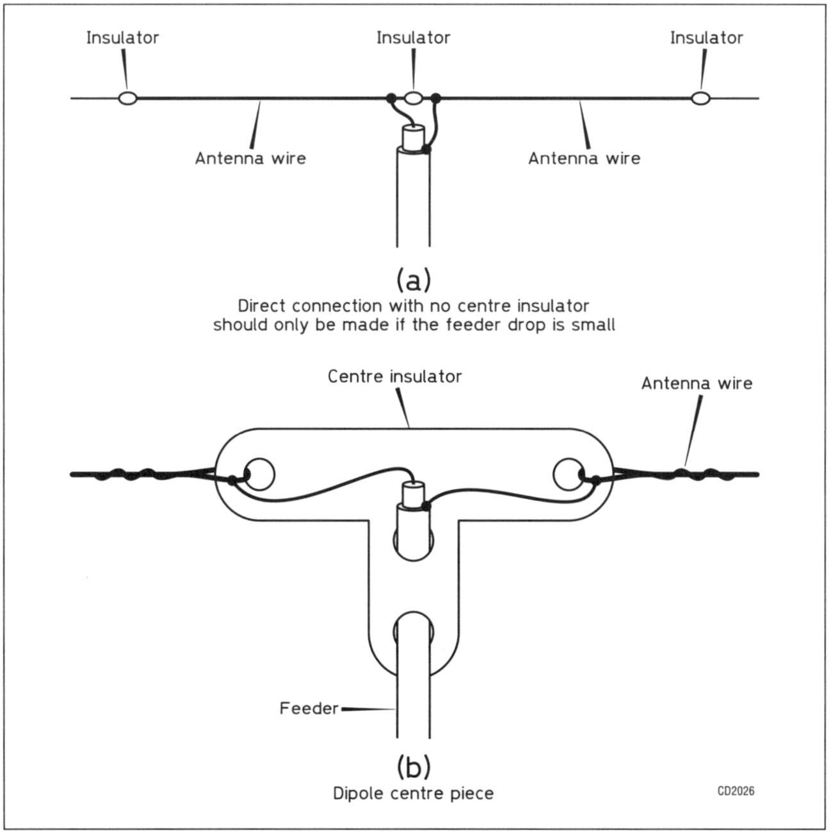

Table 6.1. Half-wave dipole lengths for the HF amateur bands

Band (MHz)	Length (ft)	Length (m)
1.8	266	82.2
3.5	137	42.2
7.0	68.5	21.1
10.1	47.5	14.7
14.00	34.3	10.6
18.068	26.6	8.2
21.00	22.8	7.04
24.89	19.3	5.94
28.00	17.1	5.28

Fig 6.9. Dipole construction

of the feeder connected to earth. Although the antenna will operate without a balun, the use of one will prevent signal being radiated from or picked up by the braid on the feeder. This may help prevent interference being caused to nearby televisions or other forms of radio equipment. The use of a balun also ensures that the normal figure-of-eight radiation pattern is preserved.

Baluns can be made or bought. In the case of feeding a dipole with 50-ohm coax they would normally be a 1:1 transformer, ie one having the same number of turns on the primary and secondary.

Sloper

It may not always be convenient to erect a horizontal dipole. There may be insufficient space or other performance issues. Fortunately it is possible to use the dipole in a number of alternative ways to enable it to be fitted into the available space.

One configuration that can be used is known as a *sloper*. It uses a single mast or tower and the dipole slopes away as shown in Fig 6.10. Sloping dipoles like this tend to have a slightly directional pattern with maximum radiation being away from the mast. For this to occur the mast must be electrically conductive because it tends to act as a reflector. If it is not conductive, for example if it is wooden, then the radiation pattern will be almost non-directional.

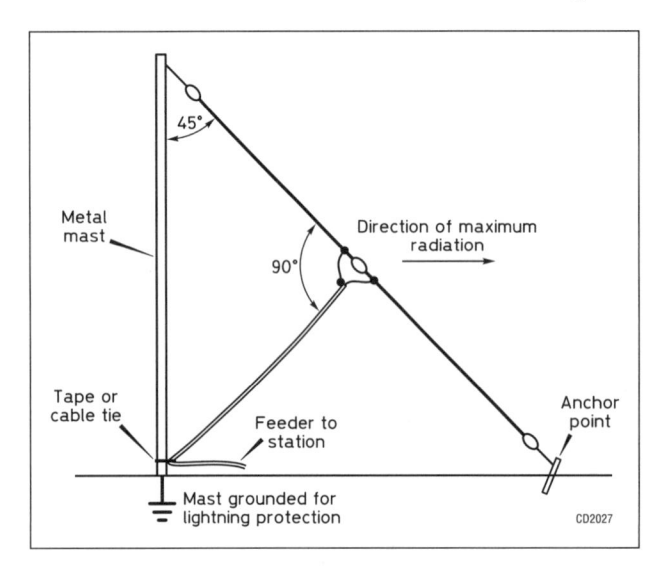

Fig 6.10. An example of a sloper. Using a metal mast makes the antenna directional – if a non-conductive mast is used the radiation pattern is almost non-directional

This type of antenna is often used by stations with a large tower which is being used to support a beam. Slopers can be arranged around the tower for other bands they may want to use. By switching in the required sloper, the direction of maximum radiation can be chosen, even if in rather an approximate fashion.

Inverted-V antenna

The inverted-V antenna is a very popular way in which a dipole can be erected. Like the sloper it only uses a single pole or mast. The feed point is at the centre, and this has the advantage that this is where most of the radiation, or signal pick-up, occurs and is at the highest point, making the most of the height of the mast. The radiation from the antenna is mainly vertically polarised, and it is also almost omni-directional.

When constructing an inverted-V it is found that the optimum lengths are slightly longer than those that might be calculated from the normal dipole formula. This is because the ends of the antenna are normally relatively close to the ground. It is difficult to estimate exactly how much longer they should be as it will depend upon the actual installation.

To accommodate this it is wise to make the legs of the dipole even longer

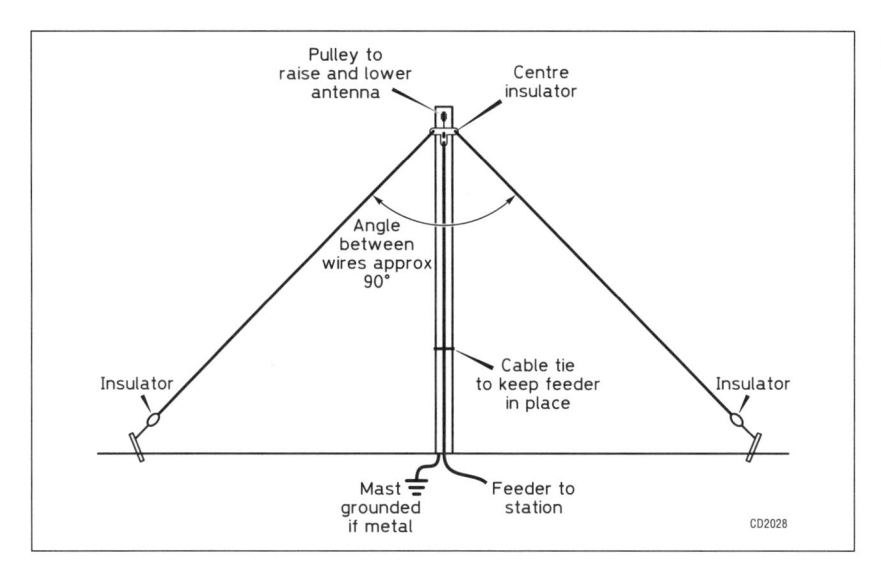

Fig 6.11. An inverted-V antenna

than normal so that they can be trimmed to size once the antenna is in place.

When deciding on the installation, the ideal angle between the two legs is 90° as shown but don't worry if this cannot be achieved, as the antenna is quite tolerant to some variation on this. The feeder should also be run along the mast to prevent it causing any unbalance to the antenna. Care should also be taken to ensure that there is no likelihood of people being able to walk into the wires as they reach the ground, trip over them or even touch them.

Once installed, inverted-V antennas can work very well and, as the wire is not horizontal, the space occupied by them is less.

Multiband dipole

One of the disadvantages of a dipole is that it is essentially a single-band antenna, although it can be used on a frequency of three times the fundamental. As most people want to be able to use several bands without having a large array of separate antennas or feeders, there are a number of ways in which this problem can be overcome.

The most straightforward is to feed several dipoles via the same feeder. When one is in operation on its resonant frequency, the others will not unduly affect its operation because they will not be on their resonant frequency.

If the wires for the different dipoles are fanned apart from one another, the interaction between them is less, resulting in less de-tuning. The reduced interaction also makes them easier to adjust to the correct length. It is also best to adjust the dipole element for the lowest band (ie the longest wire length) first and then follow through in order.

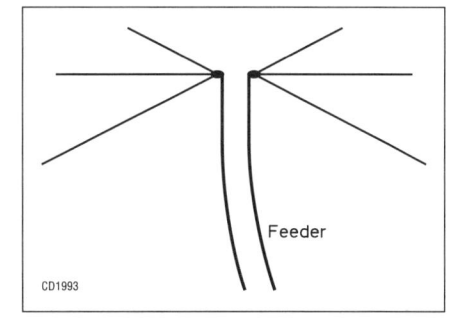

Fig 6.12. Multiband dipole

Fig 6.13. A doublet antenna serves as a versatile multiband antenna. The ATU that is used must have a connection for a balanced feeder

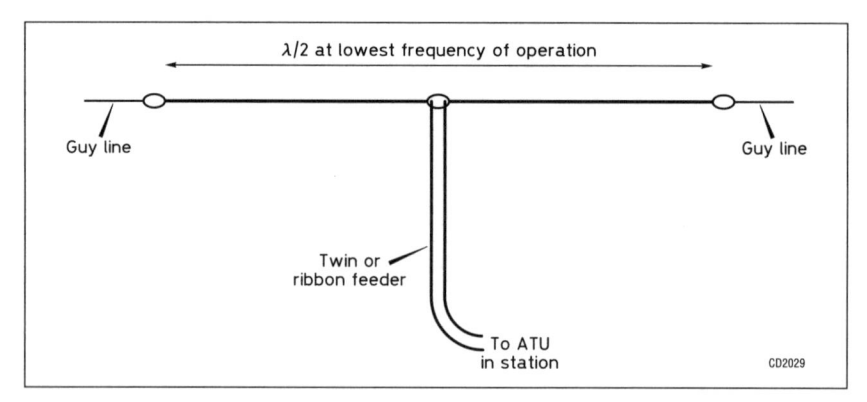

Doublet antenna

This antenna is a form of dipole but, because it is fed with open-wire feeder, it is possible to use an antenna tuning matching unit at the end of the feeder to bring the whole antenna to resonance. The length of the dipole element should be at least a half-wavelength on the lowest frequency of operation, but apart from this there are few restrictions. Changing from one band to the next simply requires the antenna tuning matching unit to be readjusted to give the optimum VSWR. Once this has been undertaken a few times the settings can be noted for a much speedier change for the future.

The main drawbacks of this type of antenna are that the tuning unit requires adjusting for each band change; the open-wire feeder should not be run through the house or close to any objects that might unbalance it; and the tuner must have a balanced output. Not all tuning units have this type of output so it is worth checking before buying a new tuner.

The advantage of this type of antenna is that it can be used for all bands above the lowest frequency determined by the length of the element corresponding to a half-wavelength at that frequency. For example, one that is 132ft (40m) long could be used on all the amateur bands from 80m through to 10m.

Trap dipole

Another popular method of creating a multiband antenna is to isolate sections of it to provide a variety of different resonant lengths. This is achieved by using circuits known as *traps*. These are parallel-tuned circuits, and at resonance they have a very high impedance. An antenna can be made to resonate on a number of different frequencies by placing one or more of these traps in each leg of it.

A typical example of an antenna using traps is shown in Fig 6.14. Here the centre part of the antenna is resonant on 14MHz because traps resonant at this frequency isolate the remainder of it when a 14MHz signal is applied. When the frequency moves away from the resonance of these traps, signals can pass through to use the rest of the antenna. The next resonant length occurs at 7MHz, and again traps resonant at this frequency prevent signals passing any further along the antenna, again isolating the outer sections of

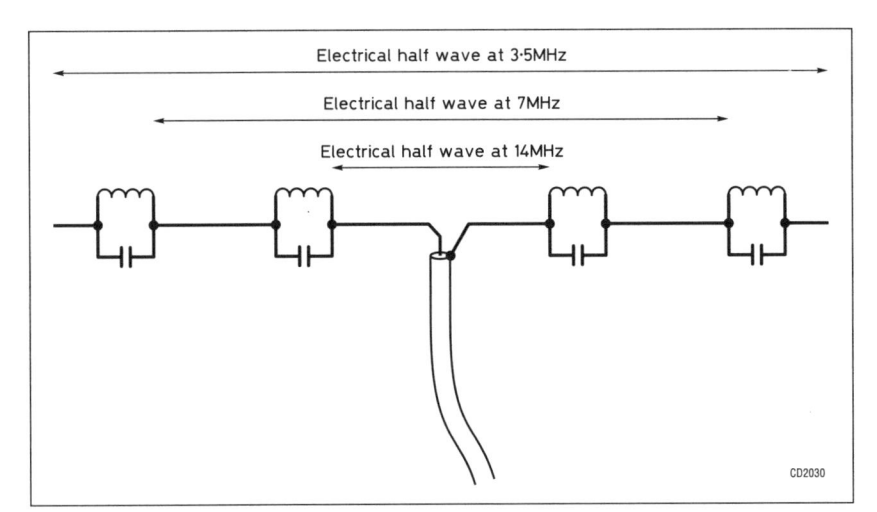

Fig 6.14. A trap dipole. The traps isolate sections at different frequencies, giving different effective lengths and allowing multiband operation

the antenna. Finally, the outermost part of the antenna is included and the whole antenna is resonant at 3.5MHz.

When using a trap dipole it should be remembered that the physical length of any paths that pass through a trap will be much shorter than their electrical lengths. This is because the trap will appear inductive below its resonant frequency and this inductance will have the effect of shortening the physical length required.

Verticals

Vertical antennas are very popular antennas, as they are able to provide very good levels of performance without requiring large areas for them to be erected. They also offer an omni-directional radiation pattern, and a low angle of radiation − a distinct advantage when wanting to contact DX stations.

A vertical antenna can be likened to half a dipole in the vertical plane. The other half of the dipole is replaced by the ground as shown in Fig 6.15. In an ideal world the ground would consist of a perfectly conducting earth but in reality a vertical will operate satisfactorily provided that a good earth connection can be made. Obviously damp soil helps to provide a good earth system.

Unfortunately rocky or sandy environments do not provide a good earth and in these circumstances a vertical antenna may not perform well. An alternative is to have a simulated earth or ground plane. This is normally constructed using a number of quarter-wave radials spreading out from the base of the antenna (see Fig 6.16). Although four radials are normally shown in diagrams, there is no set figure for how many should be used. It is possible to use only one, although performance will be slightly impaired and radiation will be best in the direction of the single radial. Often when a multiband vertical is

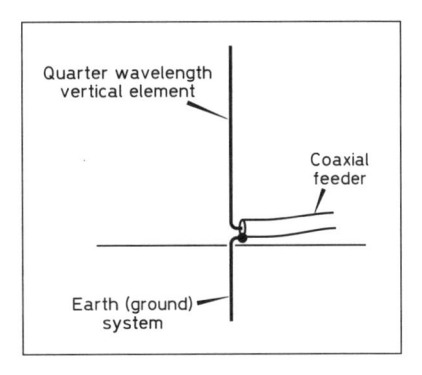

Fig 6.15. A vertical antenna

HF AMATEUR RADIO

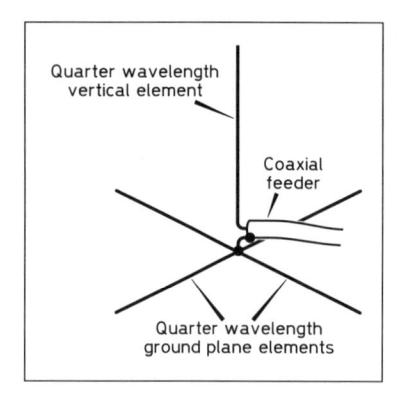

Fig 6.16. A ground-plane antenna

Fig 6.17. A Yagi antenna

used, radials of different lengths are used to ensure coverage of all the bands. Some antennas even use loaded radials to shorten their length. The advantage of using a ground-plane system is that it can be mounted above ground level, and this will improve its performance.

The angle at which the radials leave the antenna can be changed. With horizontal radials the antenna has a low impedance, but it is found that it rises if they are angled towards the ground. Often an angle of around 30° gives a good match. In this way the radials can form part of the guying system, provided that sufficiently strong wire is used. Ultimately, if the radials are angled vertically then the antenna becomes a vertical dipole and the impedance is approximately 75 ohms, dependent upon the height above ground.

Trap vertical

It has been shown that it is possible to insert traps into a dipole, and the same is also true for a vertical. In fact, trap verticals are a particularly popular form of antenna. Being able to operate on a number of bands, and not occupying much ground area, they are ideal for people wanting to operate on a variety of the DX bands from a restricted space.

Although it is possible to build a trap vertical, commercially made antennas are normally bought. There is a good selection to choose from and they are robust as they are normally made from tubular aluminium. The traps are generally made as an integral part of the antenna, being slightly wider than the rest of it. The outer sleeve forms a capacitor, while inside there is a coil.

In addition to the normal quarter-wave varieties, a number of half-wavelength antennas are available. These have the advantage that their operation is not dependent upon the earth connection, and accordingly they do not have the inconvenience of the requirement for a radial or ground system.

Beams

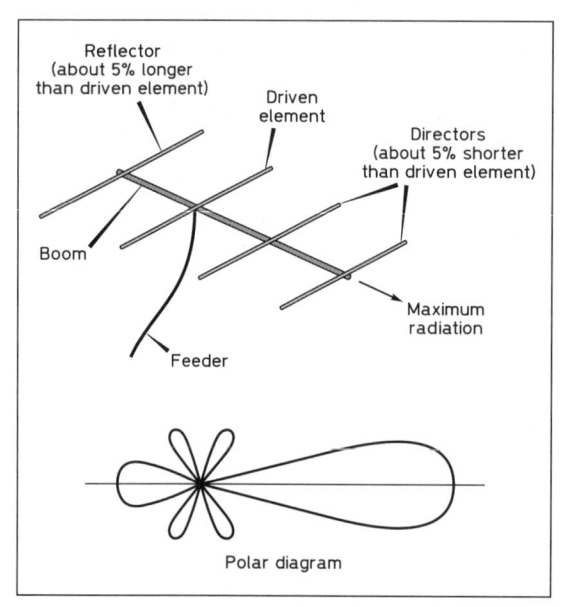

While it is not possible for everyone to install a beam antenna, many really keen DXers use them because they give improved performance over dipoles, verticals and many other types of antenna. Accordingly they give a station using a beam a distinct advantage over other stations. Not only does the antenna itself give a significant amount of gain, but also the fact that it is usually mounted on a high tower gives a further improvement.

CHAPTER 6: ANTENNAS

One of the most popular directive or beam antennas is known as the *Yagi*. Named after the Japanese researcher who invented it, this type of antenna is used in many areas of radio. Most terrestrial television receivers have a Yagi antenna connected to them, and they are also very popular with radio amateurs who need a directional antenna offering gain.

The antenna operates because placing other 'parasitic' elements close to a dipole can change its directional characteristics. These extra elements interact with the signal being radiated in such a way that the power is either enhanced or reduced in a particular direction. Elements tend either to 'reflect' or 'direct' the power and because of this they are given the names *reflector* and *director*.

A Yagi antenna is made up as shown in Fig 6.17. The reflector behind the driven element is made about 5% longer. Only one reflector is used because any further ones do not noticeably improve the performance. One or more directors are placed in front of the driven element. The one nearest it is about 5% shorter and any further directors are made slightly shorter that this.

The spacing between the elements will vary from one design to the next. It is usually between a quarter and three-eighths of a wavelength, the exact spacing being chosen to adjust the feed impedance to the required value.

An impressive antenna system using a Yagi array

A full-sized Yagi for even the highest-frequency HF bands can be quite large, and as a result it may not be possible to install one. Fortunately there are some 'minibeams' that are available. While they do not offer the same degree of performance, they may be a good alternative in many instances. Typically they have a much narrower bandwidth and may not cover a complete band. The level of gain will also be less. This may not be a significant problem, especially if the beam is well situated, possibly above the house where it will be free from obstructions and where a significant improvement will be provided by the location alone.

When using a beam, it should be remembered that the cost of the antenna itself is only part of the total cost. A rotator, possibly long lengths of

A minibeam antenna

coaxial cable, and a tower or other mounting will be required. All of these will add to the total cost of the installation, of which the antenna itself will only be a small part.

A good earth connection

A good ground or earth system is essential for the operation of some antennas. An end-fed wire (or in particular a quarter-wave, ground-mounted vertical) requires a good earth as it forms part of the antenna. The nature of the ground itself will play a big part in the way the ground system performs. If it is dry and sandy it will not be nearly as good as if the area is moist. Whatever the type of soil, a good system must be set down if the earth is to perform well.

The basic earth connection can be made by driving a long earth rod into the ground as close as possible to the antenna feed point. These rods can be bought from electrical wholesalers relatively cheaply. However, one on its own is unlikely to make a sufficiently good connection, and normally an array of several is required. In addition to earth spikes, spare lengths of copper pipe can be buried (they should not be driven into the ground as the copper is soft and will quickly bend and buckle if hammered in). Another idea that has been used is to buy an old water tank or other suitable metalwork.

Although this will provide a reasonable direct current (DC) connection, it may not give a sufficiently good radio frequency performance. To achieve this a number of *radials* or *counterpoises* may be used. These wires are laid out on the ground radiating out from the base of the antenna. Although they can be set out just above the ground, this is seldom practicable and they can be set into it instead. This can be easily done by cutting a thin slit into the ground, even through a lawn, and closing up the slit after the wire has been inserted into it. The slit will soon become invisible, even in a lawn if it is done carefully.

The radials should generally be about a quarter-wavelength long at the frequency of operation. If the antenna is to cover a variety of bands then they can be made different lengths to cover all of them. Some stations, particularly short-wave broadcast stations, have been heard to use over 100 radials. An amateur station may want to set down an 'earth mat' using multiple radials of different lengths for the bands in use. Another alternative may be to use chicken wire to cover the area around the antenna. The best rule is to use whatever you can.

Location

It has been said that the ideal location for an antenna would be on an island in the middle of a salt marsh on top of a high plateau. Even then it would help if the land fell away gently in the direction of the signal. Clearly this is not possible for the majority of radio amateurs! Most of us have relatively small spaces in which to install our antennas. However, do not be discouraged if the area available is small and far from ideal as there is a surprising amount that can be done. Even the amateurs who regularly contact stations all over the world sometimes have very small gardens.

Fig 6.18. An ideal radial earth system for a vertical antenna using many quarter-wave radials for the bands in use

There are naturally a few guidelines to follow. The first is to ensure that, whatever type of antenna is used, it is properly and safely installed. It has been said that if the antenna does not fall down in a strong wind then it is not high enough. The reply that was made to this was that a good antenna is one that is always available when it is required. Also an antenna that could fall down is a danger, and could cause damage to other's property or, worse still, injury to yourself or others. The antenna should also never be installed close to power lines – if it fell onto them it could become live and lethal.

Another point to note is that the antenna should be as far away from the sources of man-made interference as possible. Not only will this reduce the level of received interference from the many electrical items like motors, fluorescent lights, computers and so forth in the house, but it will also mean that there is less likelihood of the transmissions interfering with domestic televisions and radios.

It also helps if the antenna is not screened, particularly by buildings, because the wiring and plumbing in them will pick up and absorb the signal. If the metalwork is particularly close then it may have the effect of detuning the antenna.

If power or telephone lines run in the vicinity of the antenna, try to arrange it so that its axis is at right-angles to the lines. In this way the coupling will be reduced and pick-up of interference radiated by the latter will be minimised.

The height of the antenna is also important. In general, it should be as high as possible to enable it to 'see' over the obstructions. Also, antennas that are low only tend to transmit and receive signals that have a high angle of radiation. Low angles of radiation are best for DX use because each reflection from the ionosphere gives the maximum distance – again height gives advantages.

Safety

One major aspect associated with antennas is that of safety. Putting metalwork up into the air has a number of risks and these must be minimised as far as possible. Unfortunately it is not possible to give a complete list of all the hazards that should be avoided, but instead a general awareness of safety should be kept in mind at every stage of the installation of the antenna, from the initial concept stage right through to its operation. This is particularly relevant where large, heavy systems are installed.

In particular there are a number of areas where accidents tend to happen. Many are caused by rushing the work, possibly because someone is keen to see results or because there is a deadline such as the onset of darkness or a contest start. Under these circumstances extra care should be taken.

Work on HF antennas often involves using a ladder. This should always be securely in place, possibly with someone else at the bottom to make sure the base does not slide. It is also important to ensure that the ladder is at the right angle. When working on uneven ground, time should be taken to ensure that it is secure and that wooden or concrete slabs are used to give a firm base. Similarly stepladders should have all four feet firmly on the ground so that the ladder does not rock.

Care should be taken if mains-driven tools are required. They should not be used when it is wet, and an earth-leakage breaker should be used at all times.

When undertaking any work it is essential that someone else is around to give or call for help if the unthinkable does happen.

As for the antenna itself, there are a number of rules. First of all it should not be installed where it could fall onto any power lines. However unlikely this may seem, it has happened before and people have been killed. Similarly it is probably obvious to state that an antenna should never be erected under power lines.

The proper techniques and fittings should always be used, and the antenna should be inspected periodically. The ravages of the weather will eventually tell on even the most professionally installed system.

If in doubt it is always best to call in an expert. This is particularly important with towers and large antennas. If even a small tower blows down in the wind the results could be very serious.

If thought is given to safety there should be no accidents and the antenna can be successfully installed and used, providing many years of effective and safe use.

Further reading

Backyard Antennas, Peter Dodd, G3LDO, RSGB, 2000.

CHAPTER 7

Bands and band plans

In this chapter:

- Band plans
- Band summaries
- QRP operation

THERE is an enormous difference in the characteristics of the amateur bands within the short-wave portion of the radio spectrum. From 'Top band' in the MF region to 10m, which is almost a VHF band, there is a tremendous change in the way signals propagate, the types of antenna that are used, and in some of the circuit techniques that may be employed in the equipment. This makes the short-wave bands particularly interesting. When propagation conditions mean that one band is closed or not able to support long-distance communications, another may be open to places all over the world. So whether you are a night owl or a daytime-only person there will always be a something of interest on the bands.

One of the secrets of successful operating is knowing where to look and when. Band conditions are always changing, not only over the course of a day, but from day to day as well. Experienced operators know how to tell when conditions are likely to be good. Not only is propagation an issue, but also some of the bands are more popular than others. While this means that there are more stations to contact, it also means there is more interference and competition, so making contacts with stations in rare countries may be more difficult. By knowing which bands to use, the most efficient use can be made of the station and the time available. Band plans also differ between countries – by choosing a position in the band where there is likely to be less competition, the results may be improved. This is where the operator's skill can help make up for not having a high-power station with a large antenna on the top of a hill, and can bring success and a great sense of achievement when using an average station.

Band plans

The HF amateur bands are subject to band plans in the same way as the VHF bands. In the USA the planning is mandatory and stations are required under the terms of their licence to keep to them. In the UK and many other countries, band plans are not mandatory but adhering to them

Table 7.1. UK 160m band plan

1.810–1.838	Morse only
1.838–1.842	Digimodes
1.842–2.000	Phone (and Morse)
1.843	QRP
1.960	DF Contest beacons

makes good sense. Not only will other stations be looking for transmissions of a particular type in a given section of the band, thereby increasing your chances of making a contact, but keeping to the band plan also reduces the levels of interference and makes the best use of the available space. For example, the Morse section of the band is narrower than the section given over to single sideband because a Morse signal takes up much less space than a sideband signal. Operating a sideband signal in the Morse section of the band would cause interference to several Morse stations and accordingly this should not be done. However, it is quite permissible for a Morse signal to be transmitted in the sideband section. Although there would be little point in putting out a Morse CQ call in the sideband section of the band, it is not uncommon for a station to revert to Morse when it is not possible to maintain the contact using single sideband.

Provision is also made for other types of transmission, including data and slow-scan television (SSTV), and there are sections of the bands or spot frequencies reserved for beacons. Care should be taken to avoid transmitting in these reserved areas, particularly the beacon sections as many people will be listening to them and they will not appreciate someone transmitting on the same frequency.

Band plans are not totally the same around the world for a variety of reasons, including the fact that not all countries have the same band allocations. Summaries of the UK band plans are given below. Up-to-date, comprehensive band plans can be found on the RSGB web site and in the *RSGB Yearbook* (see 'Further reading' at the end of the chapter).

Amateur bands

Each of the amateur bands in the MF and HF portion of the radio spectrum has its own very distinct character and is likely to produce signals from different areas and at different times. By using one band instead of another at a particular time, much better results may be achieved.

160m (Top Band) (1.81–2.00MHz)

This is a challenging yet interesting band for anyone wanting to make DX contacts and one for the night owls because it only supports long-distance communications at night. However, it can be used for relatively local contacts during the day. At that time signals are heard via ground wave and, dependent upon transmitter powers and antennas, distances of 50 miles or more may be reached. At night, when the D layer disappears, distances increase and it may be possible to hear stations several hundreds of miles

Table 7.2. UK 80m band plan	
3.500–3.580	Morse
3.500–3.510	Priority for intercontinental contacts
3.500–3.560	Preferred section for contest contacts
3.560	QRP
3.580–3.620	Digimodes
3.620–3.800	Phone (and Morse)
3.600–3.650	Preferred section for phone contest contacts
3.690	QRP
3.700–3.800	Preferred section for phone contest contacts
3.730–3.740	SSTV / fax
3.775–3.800	Reserved for intercontinental contacts

away. It is even possible to make transatlantic contacts when conditions are right if sufficiently good antennas are available at both ends.

For very-long-distance contacts the whole of the path must lie in darkness. There can be a significant improvement at dawn and dusk for contacts with the other side of the globe. These enhancements may only last for 10 to 15 minutes at maximum, and sometimes less.

For shorter paths, like those between Europe and North America, signals peak when it is either sunrise or sunset at one end or the other. Long-distance, north-south paths often peak around midnight. As a general rule, long-distance work improves in winter because of the longer hours of darkness and lower levels of static. As this does not correspond with optimum conditions in the other hemisphere, it means that these signals may be heard at any time of the year.

The exact allocations for this band vary from one country to the next. Those given here are correct for the UK. Much of this band is shared with other services and this means that, when combined with the levels of static that are experienced, the levels of interference can be high.

80m (3.5–3.8MHz [3.5–4.0MHz in North America])

Like 160m this band is shared with other services and can be noisy, especially at night. However, during the day the distances that can be reached are greater than those on 160m. Often stations a few hundred miles away can be heard, making it an ideal band for contacts around the UK, for example. At night stations from further afield can be heard – distances of over 1000 miles are common, and greater distances can be achieved by those with good antennas. The band comes into its own during the years of the sunspot minimum, but it can perform well at any time.

Propagation along the grey line can produce exceedingly good results with stations from the other side of the globe being audible at the same strengths as many local stations. However, this may only be short lived and it can be very selective in terms of location.

Most of the SSB DX takes place in a 'DX window' in the top 25kHz of the European band. As a result this section of the band should be kept clear at all times. This should be observed even when you think there is no possibility

Table 7.3. UK 40m band plan

7.000–7.035	Morse only
7.030	QRP
7.035–7.045	Digimodes (and Morse – phone can be used above 7.040MHz)
7.045–7.100	Phone (and Morse)

Table 7.4. UK 30m band plan

10.100–10.140	Morse
10.106	QRP
10.140–10.150	Digimodes (and Morse)

of any DX coming through because stations with a good location and good antennas might just be able to hear DX stations and will not want to suffer from local interference. Stations in North America and other areas of the world have an allocation up to 4.0MHz so it is common to work split frequency, using the DX window below 3.8MHz for European stations and above 3.8MHz for North America etc.

Although antennas for the band can be large, very few stations have beam antennas with high levels of gain. This means that the average station can be quite competitive, especially if the full legal power limit can be achieved.

40m (7.0–7.1MHz [7.0–7.3MHz in North America])

The 40m band is a particularly useful band, providing an interesting mix of short-haul DX by day and worldwide communications at night. During the day, stations up to distances of a few hundred miles can often be heard. However, ionospheric absorption limits greater distances. The high angle of skip means that the skip zone is small or non-existent.

Distances increase considerably at night. Stations from further afield are more apparent and, as the skip zone increases, local stations fall in strength. It is a favourite band for many during the low part of the sunspot cycle, being capable of long-haul contacts during the hours of darkness. Again the grey line can produce some spectacular results.

In Europe the band is only 100kHz wide, making it congested when open for long-distance traffic. In North America, where frequencies up to 7.3MHz are available, interference from European broadcast stations (to whom this portion is allocated in Europe) can be a problem.

This band can be a good hunting ground for those with medium powers and average antennas. It is found that comparatively few people use directive antennas and this means that those with average stations are at less of a disadvantage. Trap verticals, provided they are operated against a good earth or ground-plane system, can give a good account of themselves, allowing stations all over the world to be contacted.

30m (10.100–10.150MHz)

This band was released for amateur use after the World Administrative Radio Conference held in 1979 (WARC 79). It is still not very widely used

Table 7.5. UK 20m band plan	
14.000–14.070	Morse
14.060	QRP
14.000–14.060	Preferred section for contests
14.070–14.099	Digimodes (and Morse)
14.099–14.101	Beacons only
14.101–14.112	Digimodes (and phone and Morse)
14.112–14.350	Phone (and Morse)
14.125–14.300	Preferred section for contests
14.230	SSTV
14.285	QRP

but is capable of giving good results. It is very similar in character to the 40m band, being only slightly higher in frequency.

The band is capable of giving DX contacts for most of the day, although it is generally better at night. For most of the time a skip zone is apparent, except at the peak of the sunspot cycle. The level of absorption is less than on 40m, and during the night distances increase. Again conditions are enhanced by grey line and dusk or dawn conditions. It is also found that during periods of the sunspot minimum, when ionisation levels are lower, absorption is sufficiently low to allow long-distance contacts throughout the day.

Like the 40m band, this and the other WARC bands are good bands for the DXer who does not have a really big station. Few of the common directional Yagi antennas have this band and some stations may still be using linear amplifiers that cannot operate here. As a result it means that those with more average stations will be operating at less of a disadvantage.

Due to the small size of the band and the high level of commercial activity (because it is shared with other services), most of the operation is in Morse. In fact the IARU for Region 1 have recommended that contests and phone operation should be excluded from the band.

20m (14.0–14.35MHz)

This is the main long-haul band for radio amateurs, reliably giving the possibility of long-distance contacts during all phases of the sunspot cycle. During the day, stations up to about 2000 or 3000 miles can be heard when conditions are good, and there are virtually always stations between 500 and 1500 miles which can be heard. The band normally closes at night during the winter and during the sunspot minimum, but during the summer and the sunspot maximum it will remain open most of the night. Spring and autumn normally produce good results, with stations from the other hemisphere being heard with ease at various times of the day.

Over the course of a day, signals can be heard from all over the world. In the early morning signals arrive from the east, and typically these will include signals from the other side of the globe. When these signals fade out, more local signals will become prominent, and there may be openings to the west as the Sun rises in that direction. As the afternoon wears on, openings further west may arise. There may also be openings to the other side of

HF AMATEUR RADIO

An efficient HF antenna system that could be used on 20m

the globe again as their morning approaches. In the evening, as the levels of ionisation fall, the local signals will fall in strength, leaving long-distance stations to the west.

The actual area of the openings will be heard to change. From the UK, stations in Canada and on the north-east coast of the USA are heard first in the afternoon or evening, and then the skip can be seen to move southwards, encompassing the southern states and then down into the Caribbean and South America.

Being the mainstay DX band, 20m is often crowded and, when any rare stations appear, competition is often great, making the requirement for high powers and good antennas much more imperative. Some of the big stations run powers of the order of a kilowatt (where licensing conditions permit) and three-element Yagi antennas at a height of around 60ft (20m). Nevertheless it is still possible to make many good contacts. Good operating techniques and listening for stations before the pile-ups become too large enable

Table 7.6. UK 17m band plan

18.068–18.100	Morse only
18.100–18.109	Digimodes (and Morse)
18.109–18.111	Beacons only
18.111–18.168	Phone (and Morse)

Table 7.7. UK 15m band plan

21.000–21.080	Morse
21.060	QRP
21.080–21.120	Digimodes (and Morse)
21.120–21.150	Morse only
21.149–21.151	Beacons only
21.151–21.450	Phone (and Morse)
21.285	QRP
21.340	SSTV

contacts to be made with DX stations. Often when the conditions are good it will be necessary to decide whether to stay with a pile-up or move on to find if there are any other DX stations with whom contact is more likely.

17m (18.068–18.168MHz)

Like the 30m band, this one was released for amateur use after WARC 79 and some older transceivers may not cover it. It is very much a half-way house between 15 and 20m. Although rather narrow, it is still very popular and well worth investigating when conditions look promising.

The band can offer some excellent opportunities for contacting DX stations. Although beam antennas are available for it, most stations still use dipoles as those with beams may use them for the more traditional DX bands of 10, 15 and 20m, thereby limiting the number of strong stations. However, more antennas are appearing for the WARC bands with the result that more people are using these frequencies.

15m (21.0–21.45MHz)

This band is more variable than 20m, being affected more by the state of the sunspot cycle. During the peak it is open during the day and well into the night when it will support propagation over many thousands of miles. When the band is open like this, strengths are usually better on it than 20m because of the lesser effect of the D layer. Conditions are usually not quite so good in the early morning, improving as the day progresses. During the sunspot minimum few stations may be heard during the day and none at night.

At the top of the band is the 13m broadcast band. Tuning up into this will give a quick indication of whether the amateur band may be open.

12m (24.890–24.990MHz)

This band is greatly affected by the position of the sunspot cycle and has many similarities with 10m. Although it may just support propagation when the latter cannot, it will follow very much the same pattern.

Table 7.8. UK 12m band plan

24.890–24.920	Morse
24.920–24.929	Digimodes (and Morse)
24.929–24.931	Beacons only
24.931–24.990	Phone (and Morse)

Table 7.9. UK 10m band plan

28.000–28.050	Morse
28.050–28.150	Digimodes (and Morse)
28.060	QRP Morse
28.150–28.199	Morse
28.199–28.201	Beacons only
28.201–29.200	Phone (and Morse)
28.201–28.255	Continuous duty international beacon project
28.360	QRP
28.680	SSTV
29.200–29.300	AX 25 packet (and phone and Morse)
29.300–29.500	Satellite downlinks
29.550–29.700	Phone (and Morse). Often used for FM

Like 17m this band also is quite narrow but worth investigating when conditions mean the band could be open. Also, there are few stations using beam antennas and this makes it a good hunting ground.

10m (28.0–29.7MHz)

This is the highest-frequency band in the short-wave (HF) portion of the spectrum. During the sunspot minimum it may only support ionospheric propagation via sporadic E which occurs mainly in the summer months. This gives propagation over distances of 1000 miles or so.

At the peak of the sunspot cycle it gives excellent possibilities for long-distance contacts, producing very strong signals. This band is well known for enabling stations with low powers and poor antennas to make contacts over great distances as ionospheric absorption is less than on the lower-frequency bands. In general, propagation on these frequencies requires that the path is in daylight. Despite this, at the peak of the sunspot cycle the band may remain open into the night, although it will eventually close.

There is a large number of beacons that are active on frequencies between 28.175 and 28.30MHz. In fact, many of them are outside the allocated beacon band and care should be taken not to operate on top of them.

Although much of the Morse operation is centred around the bottom of its allocation, check the upper part of this allocation, especially during contests, as levels of activity force stations further up the band. Under these conditions, competition may be less than further down the band, giving a greater possibility of making contacts.

Activity in the SSB portion of the band is often concentrated between the beacon section and 28.60MHz and a little above. However, it is again worth taking a look above this, particularly in contests.

Stations using low-power FM may be heard towards the top of the band. The recommendation is that FM activity should take place between 29.60 and 29.69MHz, with 29.60MHz as the calling frequency. There are some repeaters in the USA with outputs at 29.62, 29.64, 29.66 and 29.68MHz with inputs 100kHz lower.

QRP operation

There is a growing band of people who enjoy building and operating low-power equipment. Most of the operation is on Morse, for two reasons. The first is that the equipment for this mode is often simpler, making home construction much easier. The second is that it is possible to copy Morse at lower signal strengths than single sideband or any other form of speech transmission. This makes long-distance contacts easier to make using this mode.

QRP frequencies are reserved in each of the HF bands for QRP operation. These frequencies, which are shown in the band plans, should be avoided by high-power stations to allow those using low power to have the minimum amount of interference and hence have the best chance of making their contacts.

The definition of what QRP actually is varies somewhat. In the USA it is often considered to constitute stations operating under 100W. The UK-based G-QRP Club defines QRP as being power levels under 5W DC input. IARU Region 1 defines QRP as under 10W input and QRPp as under 1W input.

Making contacts using low powers can be particularly rewarding. However, it may not be such a disadvantage as might be first thought. Reducing the power from 400W down to 4W output represents a reduction of 20dB. A figure of 6dB is generally taken to be equivalent to an 'S' point and therefore this power reduction represents a reduction of just over three 'S' points. In other words, if a station running the full UK legal output of 400W (26dBW) was being received at S9 and it were then to reduce the power to just 4W (6dBW), it would still be copied at around S6. While a QRP station might not be able to operate through many pile-ups, the figures show that it should still be possible to make plenty of contacts.

Along with using low power, QRP operators usually take pride in having built much of their equipment themselves. This also brings a great sense of achievement when contacts are made. Some people enjoy making their equipment by buying the components and making the equipment from scratch, while others use some of the growing number of kits that are available. Some of these are relatively simple but there are a number of full single sideband transceivers available.

Further reading

Amateur Radio Operating Manual, 5th edn, ed Ray Eckersley, G4FTJ, RSGB, 2000.

RSGB Yearbook (published annually by the RSGB).

CHAPTER 8

On the bands

In this chapter:

- Starting out
- Basic contacts
- DX techniques
- Getting through in a pile-up
- QSL cards
- Awards
- Contests

OPERATING skills are important to anyone using the HF bands. Knowing where to listen, when, and how best to make contacts can make a great difference to the types of station that can be contacted, and often a much greater difference than the equipment itself. Developing good operating skills will also mean that you are less likely to cause interference to others and you are able to conduct contacts more effectively, especially when conditions are poor. These skills take a little while to learn but it is possible to pick them up quite quickly. Even experienced operators who can handle pile-ups confidently had to start somewhere!

Starting out

To find out about operation on the HF bands it is worth taking some time to listen to them. Indeed, one of the best apprenticeships for anyone wanting to operate on the HF bands is to spend some time as a short-wave listener. In this way it is possible to find out how typical contacts are made, what is said and how the information is put over in a concise way so that the other station is totally aware of what is happening, even when there are high levels of interference – it is essential to know how to handle a contact under difficult conditions. However, to start out it is useful to know the basic format for a contact.

Basic contacts

Many of the contacts that take place on the short-wave bands are what is termed 'rubber stamp' contacts. These are a good starting place for many people, including those who do not speak English as their first language as it is relatively easy to have a contact using a minimum vocabulary. However, many people like to talk about far more than the topics covered by the basic contact. Often they will discuss technical matters or describe the part of the world where they live. Many will want to have long conversations once they are familiar with operating but first of all they need to know the basic elements of a contact so they can get started.

On the HF bands a contact will often commence with a *CQ call* (a general call to all stations). A formula known as 'three times three' is a good starting point. Here the letters 'CQ 'are repeated three times, and then the callsign, usually spoken using phonetics, is repeated three times. This whole procedure is repeated three times. In this way the call is kept to a reasonable length and anyone listening is able to gain the callsign and the fact that a contact is wanted.

Any station listening who wants a contact can then respond when invited to do so. He will normally give his callsign a couple of times using phonetics and then invite the other station to transmit.

If the first station hears the caller and responds, he will announce the callsigns and then normally wish him 'good day' and give a signal report. This tells the other station how he is being received and gives information about his station's performance, so that if conditions are difficult, the contact can be suitably tailored. After he gives the report it is normal for him to give his name and location. Then the callsigns will be given and transmission handed over.

The second station will follow a similar format, giving a report, his name and location. On the next transmission, information about equipment in the station – the transmitter, receiver or transceiver and the antenna – is often given. Details of the weather are also often mentioned.

On the third transmission details about exchanging QSL cards may be given and then the two stations may sign off. Again, callsigns will be given at the beginning and end of each transmission.

Once a contact has finished it is perfectly permissible to call one of the stations. Normally the frequency 'belongs' to the person who called CQ initially, but if the other station is called he may ask to keep the frequency or move off to another one.

Giving callsigns at the beginning and end of each transmission may seem somewhat formal, but it fulfils legal requirements to identify the station and also serves to let the other station know exactly what is happening and when he is expected to transmit. Using good operating technique is very important and helps contact to be maintained with the minimum possibility of confusion, especially when conditions are poor or interference levels are high. In essence one of the keys to becoming a good operator is to let the other station know exactly what you are doing and not leave him guessing.

When a station in a very rare location is on the band, contacts are usually kept very much shorter to enable as many as possible to be made, and this also applies to contest operating. Usually the contact will consist of just the callsigns of the stations and then a report. Speedy operating is of the essence

> **A typical Morse contact**
>
> CQ CQ CQ, DE G3YWX G3YWX G3YWX CQ CQ CQ, DE G3YWX G3YWX G3YWX CQ CQ CQ, DE G3YWX G3YWX G3YWX AR K
>
> G3YWX DE G3XDV G3XDV AR KN
>
> G3XDV DE G3YWX GM OM ES TNX FER CALL UR RST 599 599 = NAME ERE IS IAN IAN ES QTH STAINES STAINES = SO HW CPI? AR G3XDV DE G3WX KN
>
> G3YWX DE G3XDV FB OM ES TNX FER RPRT UR RST 599 599 = NAME IS MIKE MIKE ES QTH NR LONDON LONDON = SO HW? AR G3YWX DE G3XDV KN
>
> G3XDV DE G3YWX FB MIKE ES TNX FER RPRT = TX ERE RNG 30 WATTS ES ANT VERT = WX FB SUNNY ES ABT 23 C = SO HW CPI? AR G3XDV DE G3YWX
>
> G3YWX DE G3XDV R R AGN IAN = RIG ERE RNG 100 WATTS ES ANT DIPOLE UP 10 METRES = WX WET ES COLD ABT 15 C = G3YWX DE G3XDV KN
>
> G3XDV DE G3YWX FB MIKE ES UR RIG DOING FB = ERE QRU = QSL VIA BURO = 73 ES HPE CUAGN SN AR G3XDV DE G3YWX KN
>
> G3YWX DE G3XDV R R QRU ALSO = QSL FB VIA BURO = SO TNX FER QSO 73 ES BCNU AR G3YWX DE G3XDV VA
>
> G3XDV DE G3YWX FM 73 ES BCNU AR G3XDV DE G3YWX VA
>
> Note: the '=' sign is used as a full stop or break; procedural characters AR, KN and VA are sent as a single character.

under these conditions to ensure that others are not kept waiting and as many people as possible are able to make a contact.

Similar contacts can be made when using Morse, the main difference being that far more abbreviations are used to ensure that the speed at which information can be passed is as quick as possible.

Tricks of the trade

When some leading DX operators were asked what the key to their success was, virtually all of them said it was listening. While there is a great temptation to put out a CQ call and expect the rare and exotic DX station to reply, this is seldom the case. Instead, listen around the bands to find out what is going on, so you can find out if there is anything of interest on the band and see what conditions are like. If you spend all the time transmitting then you can't be searching out the interesting DX.

Even when you spend plenty of time listening there are a number of easy ways to help find the more interesting stations faster. Listen to find out if the operator is speaking with a different accent to the rest of the stations on the band, or if he sounds different in some way. If so he may be from a different part of the world. Also, when tuning up and down the band it is worth listening out for a 'pile-up' where many stations are trying to call a particular station. The station under the pile-up is almost certainly going to be of interest.

Also listen out for signals that sound different in other ways. For example, those that come across the polar regions often have a 'watery' or 'fluttery' sound to them, indicating they are some distance away. It is also possible to pick up some interesting stations as the band is closing at night. At this time interference levels may be less from short-skip stations and competition from stations to the east will also be less. Furthermore, long-distance stations are often heard and there is less competition in being able to contact them.

Information is essential

While the greatest weapon in the DXer's armoury is listening, information is also vitally important. This can be general knowledge, possibly about technical issues, or it may be up-to-the-minute information about what is going on.

A good knowledge of how signals propagate on each band and the way in which they vary over the day is vital. These conditions also vary from day to day. Predictions about the conditions on the bands can be ascertained from magazines like *RadCom* as well as information available on the Internet (see Chapter 2 – 'Propagation'). Details of the A and K indices can be picked up and judgements made about the likely conditions. While propagation predictions are much better these days, it is always valuable to listen around the bands to see if they are as predicted. Don't just check the one that is most likely to be the best, but others as well.

It also helps to know which stations are likely to be on the bands. Information or 'intelligence' about the activity of DXpeditions and other DX stations, such as their operating times and frequencies, is all useful. For non-DXpedition stations the routine of daily life often means that they tend to have the same time each day for amateur radio operating and they may also have their favourite frequencies. It has been said that DXers have an insatiable appetite for information (and even rumour) about DX stations. There are a number of ways in which this can be gained, and one of the most obvious is from magazines like *RadCom*. Every month there is an HF column giving details of the DX operations that are on the bands and information about forthcoming operations. This is very useful for giving advance warning about what is going on.

The Internet is also a very good source of information with a number of useful web sites. The RSGB's own web site (www.rsgb.org) gives some DX news but also some links to some excellent web sites, including some that give DX information. These sites are worth adding these to the favourites list on your browser. Another useful site with links is www.radio-electronics.com.

One particularly good method of quickly finding out what has been happening on the bands is by using packet radio. There is a system known as the *DX Packet Cluster* which enables people to see what DX is or has been on the bands and on what frequencies. Its advantage is that it is almost real-time. Some years ago, Dick Newell, AK1A, developed a form of real-time electronic conferencing using packet radio which allowed all the stations that were connected to see everyone else's messages. The idea was taken up by the DX community who saw the advantages the system could offer for passing information around quickly about any DX stations that had appeared on the bands or other relevant news. Now the DX Packet Cluster has become one of the most useful sources of real-time information available to the DXer. There are also some DX Cluster web sites.

The network of stations in the cluster is growing and is present in most countries that have a significant amateur population. For example most of England is covered and there are links to similar clusters on mainland Europe. There are also clusters in the USA.

Once information is entered it rapidly spreads throughout the system and can be seen by every station connected to it. The system also caters for short one-line announcement messages, 'talk through' to individual stations, conventional store-and-forward mail and bulletins, WWV propagation data, as well as access to several on-line databases. The system stores data so it is possible to log in and find out what has been happening over the past few hours. Information can be sorted and this can help in looking for patterns of operation etc which might help in predicting the best time to look for a particular station or country.

All that is needed to use the DX Cluster is an ordinary packet station which is connected with the nearest cluster node, direct or via a conventional packet node. Packet stations can be set up quite easily these days, possibly using equipment that is already available in the shack – this makes it a very attractive and efficient way of finding out what is happening on the bands.

Often DXers will alert one another to rare DX by telephone. In fact some have managed to enlist the help of local short-wave listeners so that if they hear anything of interest they quickly give a call. Telephone calls are particularly useful because they give almost instantaneous alerts and enable the unexpected appearance of a rare station to be captured.

Another useful source of information can be some of the DX nets that are organised. These change from time to time, so they are not included here, but details do appear periodically in the magazines' HF columns. It may be best to listen and hear all the latest information, or you may want to join in and share your experiences and information.

Pile-ups

When listening around the bands a pile-up is an almost sure-fire indicator that there is an interesting station beneath it. Pile-ups require a little extra operating skill to give the best chance of getting through without causing interference to others. Again, the best advice that can be given is to listen. Find out the callsign of the station, listen to hear what the station sounds like, so that he can be identified easily with all the other stations calling. Find out his mode of operation. By doing this you will know the best time to call.

Many very rare stations operate on *split frequencies*. The reason for this is that when the pile-ups get very large the DX station needs to be heard by all the others so that he can control the pile-up without being swamped by interference from those trying to call him. Typically SSB stations may listen 10 and 20kHz high in frequency or they may listen over a band of frequencies. Those using Morse are often about 2kHz from their transmit frequency.

Screen shot of a typical contest contact logging software package

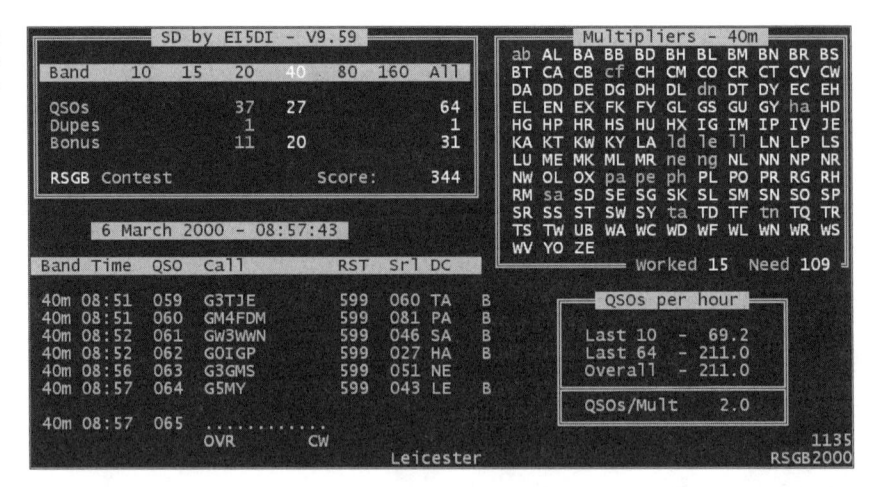

The other advantage of this approach is that it can spread the stations out, making it easier to pick out a single station from the enormous sound cacophony of the pile-up. When this mode of operation is employed, those stations who have two VFOs on their transceiver can listen to the stations being contacted and follow the tuning pattern for the DX station, and thereby they know the best frequencies to use.

When calling a station in a pile-up it is also necessary to get one's call in at exactly the right time after the previous contact has finished and when the DX station is listening. Quick and exact operating is the name of the game, giving one's callsign quickly and clearly. However, in the excitement it is necessary to be careful not to call too soon or out of place, thereby causing interference.

Some pile-ups may be so big that it is just not worth calling, as you may spend many hours calling to no avail. Instead it may be worth making a note of the frequency and then looking for other DX stations on the band, returning later when propagation conditions may be more favourable. Remember, though, it is often worth making a few calls just in case it is possible to get through. Some of the most satisfying contacts come when you did not expect to get through or when you have been rewarded after spending a long while calling.

Automatic logging programs

Many of the leading DX operators use computer-based logging software packages. While some people still like to use more traditional paper-based logbooks, there are many advantages to using a computer-based system. Not only is it possible to make duplicate copies of a log, for example for sending off after a contest, but it is possible to check on countries contacted, confirmed and so forth. In fact, with computers being very good at data handling, these packages can provide a great number of statistics very easily. On top of this many packages are able to provide partial automatic operation on Morse, sending messages like 'CQ' and the standard replies, making it much easier to operate for long periods of time. They can even be

A typical QSL card

used for voice operation via PC sound cards, if only for calling CQ. There are a number of packages that are available – some are more suited to home operation, whereas others are aimed at contest operators. These packages are continually being updated so before buying one of them it is worth talking to people who use them to gain the latest information about the best one to buy.

QSL card collecting

Many people enjoy confirming the contact with a QSL card once a contact has been completed. While few send cards for every contact these days, the volume passing through the QSL bureaux is still very high. The RSGB one alone processes around 1.5 million cards a year. Once received, they can be displayed on the shack wall to show what DX has been worked, or collected so that they can be used to prove contacts have been made when applying for operating awards.

There is an enormous variety in the types of cards that are sent. Some are printed in a single colour, although many have are multi-coloured or have photos on them. Obviously these cost considerably more to print, but with modern printing processes being used they are not as costly as they used to be.

Although there is no hard and fast rule about how a card should look or exactly how the information should be presented, it should contain certain details about the contact. The card should obviously have the callsign of the station printed prominently on it. The address or location should also be included along with the name of the owner of the station. Space is then provided to fill in details of the particular contact, including the callsign of the station with whom the contact was made, the date and time (usually in GMT or z). The frequency band or frequency as well as the mode in use is

HF AMATEUR RADIO

A selection of QSL cards from around the world

required. The report that was given in the contact should be included. Details of the equipment are very useful. Finally there is normally space to give the details of whether a return QSL card is wanted, and the route that can be used. Typically the wording is "PSE / TNX QSL DIRECT / VIA BURO". Finally there is a space for the operator's signature.

Most people will send their QSL cards via the bureau. This is by far the cheapest option but return cards can take months or years before they are received. Cards can be sent directly for those chasing awards or wanting to have a rare station or country confirmed. When sending cards directly it is normal to send the return postage. For foreign destinations dollar bills can be used, although the normal way is to use International Reply Coupons which can be exchanged for return-rate, surface-mail postage. These can be bought from main post offices in the UK but not from sub Post Offices which are unlikely to even know what they are! Unfortunately they are rather expensive and as a result many people do not cash them for return postage, instead either selling them on at a cheaper rate, or re-using them for a card they want. This is perfectly acceptable as they do not have an expiry date and as a result they have almost become an international form of currency.

One of the major problems in sending cards directly is that of finding the address of the station. Address listings of radio amateurs throughout the world can be bought from the RSGB on CD-ROM. Alternatively there are a number of web sites where it is possible to obtain this data (eg www.qrz.com), although addresses may not be completely up to date.

A number of stations use QSL managers to handle their cards for them. A station on a remote island or another inaccessible location may not have a reliable or frequent postal service. Accordingly it is far more convenient to send logs to someone in a country with all the required facilities to act as the manager. DXpeditions also need a person to whom cards can be sent, as the operation will be over once the cards arrive. Details of QSL managers are usually mentioned occasionally during the operation, so it is worth listening out for this information. Additionally, details are published in magazines such as *RadCom* and on DX web sites.

Awards

Another interest that many people have is working towards gaining some of the many operating awards that are available. These present a challenge, and once the awards have been received they look very attractive and can be mounted in picture frames and displayed in the shack.

There is a wide variety of awards that can be obtained but one of the most famous is the *DX Century Club Award.* This is awarded by the American Radio Relay League (the US national amateur radio society) to people who can supply proof (QSL cards) that they have made contact with stations in 100 'entities' (this usually means countries but see the rules). For people who have contacted more, there are endorsements that can be added. In fact some people have made contact with over 300 entities, and this represents a considerable achievement. A summary of some of the major awards is given in Table 8.1.

Contests

Contests can be an excellent way of contacting rare stations in new countries. During the major contests the bands come alive with stations, and many individual people and groups travel to different countries or islands to activate a more-sought-after country or prefix. Furthermore, some stations

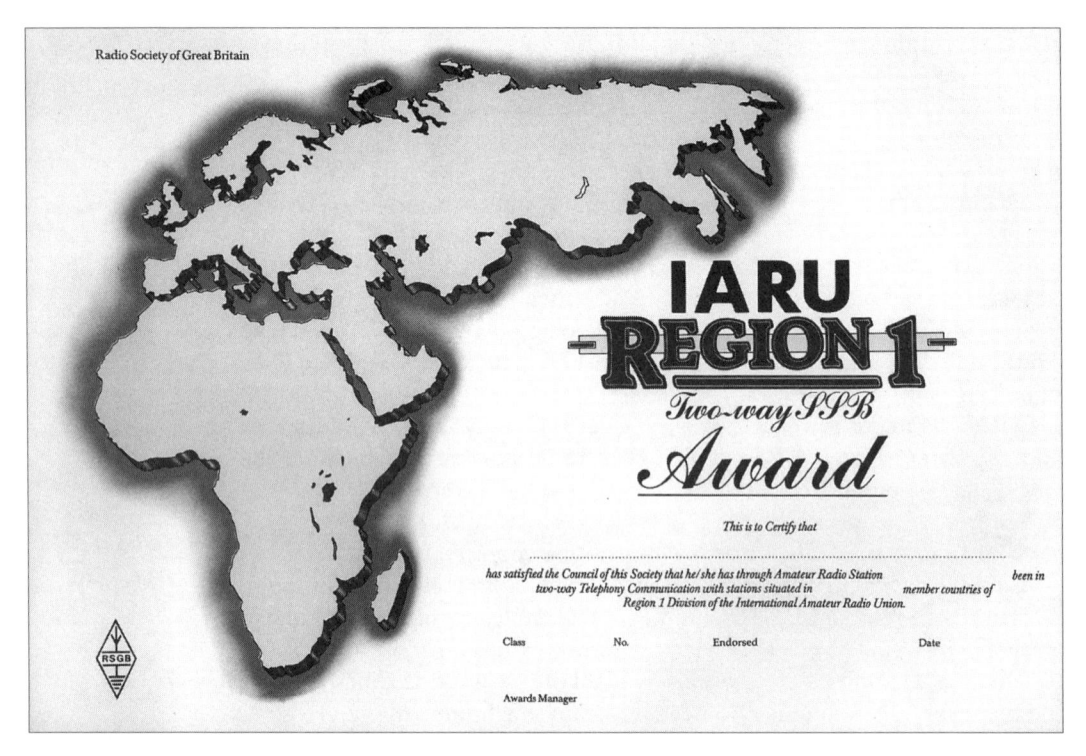

A typical operating award

in rare countries come on the air just for the contest. Coupled with the fact that contest contacts are very fast, so that as many stations as possible can be contacted, it all makes an ideal hunting ground for rare and interesting stations. In fact it is by no means uncommon for people to contact 100 countries and more during a contest weekend.

There are many contests during the year, each one being different to the others, although they all offer opportunities to make a good number of contacts and can also be great fun to enter. Some of the contests are naturally larger than others – the main ones include the ARRL DX Contest, CQ WPX, CQ Europe, All Asia DX Contest, IOTA and the CQ World Wide Contest. Of these possibly the CQ World Wide is the most popular, and there are generally a good number of DXpedition stations active specially over the weekend of the contest. Also, for each of these contests there is a phone or SSB leg, and a Morse leg on different weekends, often about a month apart.

When operating in a contest be prepared to make contacts very quickly. Everyone wants to make as many contacts as possible and there is no time for exchanging pleasantries. I find it helpful to have the logbook in a position where entries can be made as quickly as possible and to have some spare paper handy for jotting down any callsigns. Others prefer to use a computer logging system.

If you have not operated in a contest before, take a while to listen to the contacts being made to find out how it works. Normally the station who 'owns' the frequency will put out a short CQ call and wait for replies. When

Table 8.1. Summary of some of the major HF operating awards

Award	Organisation from which rules can be obtained	Notes
DX Century Club (DXCC)	ARRL	Awarded for submitting proof of contact with at least 100 DXCC 'entities'. Endorsements are available for further entities. There are several categories: Mixed (ie any mode); Phone; CW; RTTY; 160m; 80m; 40m; 6m; 2m; Satellite; Five band DXCC (100 entities on each of five bands. There is an 'honor roll' for those who have contacted within 10 of the maximum number of entities available at any given time.
IARU Region 1 Award	Can be obtained via RSGB	This may be claimed for producing evidence of having made contact with stations in countries that are members of Region 1 Division of the International Amateur Radio Union. There are three classes of the award: Class 1 for contacts with all member countries; Class 2 for 45 member countries; Class 3 for 30 member countries.
Islands on the Air (IOTA)	RSGB	This is awarded for providing proof of contact with stations located on islands (worldwide and regional). The basic award is for 100 islands, but higher achievement awards are available for 200, 300, 400, 500, 600 or 700 islands. Currently over 700 are listed.
Worked All Britain	Worked All Britain Group	This award is available for contacts (it is also available on a 'heard' basis for short-wave listeners) with UK National Grid 10km squares which have references consisting of two letters followed by two numbers, eg TQ99. The award is given in different categories (basic, bronze, silver etc) dependent upon the number of squares contacted or heard.
Worked All Continents	Can be obtained via RSGB	This is issued by IARU headquarters for providing evidence of contacts with amateur radio stations in each of the six continents – North America, South America, Europe, Africa, Asia and Oceania. For UK stations cards may be sent to the RSGB HF Awards manager who will supply a certified claim to IARU.
Worked All Zones	CQ Magazine	This is awarded for confirmed contacts with stations in each of the 40 CQ Zones. A five-band version is also available.
Worked ITU Zones	Can be obtained via RSGB	This award is available for contacts with states in 70 of the 75 ITU Zones.

he hears a station he will mention his callsign and give a report and contest serial number. The other station will respond by confirming reception of the report and serial number by saying "roger" or "QSL" and then give a report and serial number back. Finally, the first station will confirm reception and say "73s and QRZ".

Table 8.2. Major amateur radio HF contests

Contest	Date	Comments
ARRL DX Contest (CW)	Third full w/e February	Stations contact USA/Canada
ARRL DX Contest (SSB)	First full w/e March	Stations contact USA/Canada
CQ-Worked PrefiXes (WPX) (SSB)	Last full w/e March	Stations contact as many stations as possible. Extra points given for new prefixes contacted.
CQ-Worked PrefiXes (WPX) (CW)	Last full w/e May	Stations contact as many other stations as possible. Extra points are given for new prefixes that are contacted.
CW Field Day (UK) (CW)	Usually 1st w/e June	British portable stations make as many contacts as possible.
All Asia (SSB)	Third full w/e June	Contact stations in Asia.
IARU-Radiosport (CW/SSB)	Second full w/e July	Contact as many stations as possible. Extra points given for new countries contacted.
RSGB IOTA Contest (CW/SSB)	Last full w/e in July	Contact as many stations on islands as possible (note that the UK mainland is an island)
Worked All Europe-DX (CW)	Second full w/e August	Stations outside Europe contact as many European stations as possible.
All Asia (CW)	Last full w/e August	Contact stations in Asia.
SSB Field Day (SSB)	First full w/e September	Portable stations make as many contacts as possible.
Worked All Europe-DX (SSB)	2nd full w/e September	Stations outside Europe to contact as many European stations as possible.
CQ-WorldWide (SSB)	Last full w/e October	Contact as many stations in as many countries as possible.
CQ-WorldWide (CW)	Last full w/e November	Contact as many stations in as many countries as possible.

In most contests the idea is to contact as many other stations as possible. However, points may be gained in a number of ways. They may be given for each station contacted, but there may be more if they are in another country of continent, and then multipliers may be given for the number of different countries (or zones etc) contacted. Each contest has its own rules and they differ from one to the next.

Contest report numbers also vary. In some contests they may be the serial number of the contact, starting from 001, or in others it may be the zone in which the station is located, while in others it may be the power. In one contest the operator's age is sent (ladies send 00!). A summary of the major contests is given in Table 8.2.

It is also worth having a good look around the bands during a contest. Sometimes those that may normally appear closed will suddenly appear to support communication. The reasons for this are two-fold. First, many of the 'big' stations with large antennas come on the bands and can be heard.

Second, more people stay up around the clock to make contacts whereas at other times of the year they would not be active from that particular region at that time of day or night. It is therefore worth spending some time assessing what activity levels are like.

Part of the skill of operating in a contest is knowing when to stick at contacting a station and when to move on. Competition is usually fiercest at the beginning. However, towards the end, for example on the second day of a two-day contest, the pile-ups may be smaller as many people will have made contact with the rare stations. Accordingly Sunday afternoon and evening can be very rewarding times on the bands, although some pile-ups can still be large and difficult to get through.

Further reading

Amateur Radio Operating Manual, 5th edn, ed Ray Eckersley, G4FTJ, RSGB, 2000.

RSGB Yearbook (published annually by the RSGB).

Packet Radio Primer, 2nd edn, Dave Comber, G8UYZ, and Martyn Croft, G8NZU, RSGB, 1995.

CHAPTER 9

Setting up the radio station

In this chapter:

- Requirements for the shack
- Ideas for locations for the shack
- Constructing a table
- Mains distribution
- Lighting
- Choosing equipment
- Equipment layout
- Decorating the shack
- Safety

IT IS important to have a radio area ('shack') that is well planned. Since a lot of time is likely to be spent there, listening, transmitting, constructing or doing any one of a number of tasks, it is necessary that the room is easy to use and comfortable. If it is difficult to use then the level of enjoyment will be reduced considerably, but if it is a well-set-out shack it will bring more enjoyment and encourage more use of the equipment and space.

The first step in setting up a shack is to determine where it will be. This is not always as easy as it might seem. While the ideal location would be a spare room, this is not always feasible. Fortunately it is possible to set up a shack in many different places around the home by using a little imagination and ingenuity – walk-in cupboards, garden sheds, backs of garages and a whole host of other places can be used.

Shack requirements

Everyone will have their own ideas about how they want their shack to be set out. Some people will want to use it primarily for operating, while others will want the space for construction, and some will want it to cater for both. This makes hard and fast rules about the ideal shack difficult to lay down. The only principle that can be applied to every shack is that it pays to put in some thought and planning before actually setting it up.

During these initial stages of planning there are a number of common points that should be considered. Features to be investigated should include the amount of room required, availability of power and access for feeders.

Another point that may be of importance is the possibility of shutting the shack off from the rest of the house. If radio equipment is located in the living area it will certainly lead to comments from the rest of the family about the noise. On top of this it is not wise to leave the equipment open to others, particularly if there are likely to be children around, because their fingers seem to get everywhere!

Even though it should be possible to shut the shack off from the rest of the house it is still a good idea to have it relatively accessible. This is because it is often nice to spend just a few minutes in there to see if the 15m band is open, for example, or to find out if that rare DX station is around. If the shack involves much effort in getting to it for one reason or another there is a tendency not to bother to use it and miss out on the wanted DX.

Another factor to consider is the necessity of getting feeders in and out. Although coaxial cables can be run around the house without too much difficulty they can look rather unsightly. In view of this the shack should have ample access to the outside world. This becomes even more important if open-wire feeders are to be used at any time. They certainly do not appreciate being run around the house.

Size is another important feature. In most cases the actual operating space is only part of what is required, the rest being taken up with components, surplus equipment which may (or may not) come in useful one day, books and all the rest of the paraphernalia which goes with every amateur station. This means that the shack has to be large enough to accommodate not only the equipment itself, but also the rest of the trappings, as well as leaving some space to actually get in.

What's available

After deciding on the requirements for the shack the next stage is to look at what is available. Unfortunately the various places in which a shack could be set up seldom live up to what one would like. This is where a little ingenuity comes in useful to make the best out of what is available.

One possibility for siting a shack could be a large cupboard. At first sight this might not seem to be a particularly good solution because of the space limitations. However, if it is well planned, making the best use of all the space that is available, then it can prove to be quite an acceptable solution. A table top can be built in to the cupboard, but remember to leave sufficient space behind the surface to feed up cables without having to remove the plugs. Also remember that access for power and antenna feeders will be required without undue intrusion into the rest of the home.

Another possibility for a shack may be an outside shed. These are not always ideal because they can become very cold in winter, they require power to be installed and security may be a problem, but they do have a number of advantages. Antennas can often be located quite close by, the radio equipment is kept out of the house, and they can provide a good self-contained room for all the equipment. If this option is taken up then it is well worth lining the shed to ensure it remains warm in the winter and it looks more comfortable.

CHAPTER 9: SETTING UP THE RADIO STATION

G4BWP keeps all of his equipment within easy reach

Some people have used a convenient corner in a garage. In many ways this is not ideal, but is may be possible to convert an area of the garage, and even partition it off from the rest of the space so that it becomes warmer and more comfortable.

For many people one of the most logical and convenient places to set up their shack will be in the loft or attic. Having the shack there has several advantages. For example, a loft can often be reasonably easy to convert it into a shack, and it will usually be fairly spacious, giving plenty of storage room. Another advantage is that the shack will be separate from the rest of the house, while access to it is reasonably easy.

However, there are a few disadvantages that should be considered before making the final decision. Probably the first and most obvious question to ask is whether or not the loft timbers will stand the weight of all the equipment and people who may go up there. If there is any doubt over this it is worth consulting a builder or surveyor for his opinion. The cost will also have to be considered – a floor will have to be put down, a loft ladder installed as well as having to install power and any other work that may be needed. Also, remember that a loft will suffer from large variations in temperature – in summer it will become very hot and in winter it can be very cold.

Unfortunately there are only a comparatively small number of people who possess a spare room that can be devoted to amateur radio. However, if one can be used then it can be a very convenient option to choose. Not only will it be warmer and more comfortable, but it will be much easier to spend the odd five minutes in there as it will not necessitate a trip down the garden to the shed, going out into the cold garage or whatever. On top of all this it is likely to require less work in getting the shack set up. There will be no problems of having to fit a floor as in a loft, or line the walls as in a

G4CAO operates from what used to be the 'smallest room'

shed. Also there should be sufficient power already installed which means that there may be no necessity to alter the mains supply.

Making an operating table

Wherever the shack is located, in a loft, spare room, shed or anywhere else, some form of table will be required. One solution is to look in the shops for a sturdy desk. This can either be obtained new, but a much cheaper option is to look in a second-hand or surplus office furniture shop. Computer tables can also make stylish operating tables.

If it is not possible to obtain exactly what is required, then it is not difficult to build a suitable table. It is possible to make a good job if the design is kept simple and care taken during its construction, even for those with limited woodworking skills.

There are a variety of ways in which a table can be easily constructed. The main requirements are that it should be large enough and sufficiently sturdy. Even with modern equipment the weight soon builds up if a few pieces of equipment are placed upon it. The first step in the construction is to set the size. The depth should be sufficient to allow space behind the equipment for cables, remembering coaxial cable has a finite bending circle, and normally about 3in should be sufficient. Then there should be adequate space in front of the equipment to allow easy operation of the equipment – allow around 15in for this as a guide. This will give space for a log book, microphone, Morse key and room to rest your arm to make for easy operating during long contests.

The work surface can be made from a sheet of blockboard or plywood, although chipboard or MDF can be used if is of sufficient thickness and density. The table top can be supported by a framework of 2in by 1in wood to give extra support and prevent any tendency to sag. This can sometimes be a problem, especially if heavy equipment remains on the table for a long period of time. This framework can be fixed to the table top using countersunk screws as shown in Fig 9.1. Although this does leave holes in the work top they can be filled and then the whole surface can be covered with Formica.

The framework enables the legs to be attached more easily, in addition to providing extra support for the table top. There are a variety of ways in which this can be done. Wooden legs can be attached to the framework, but remember to have 'struts' as shown to provide support at the bottom of the legs. There are many other ingenious ways of making suitable legs for the table. For example, it may be possible to use the steel legs from an office desk that is beyond repair.

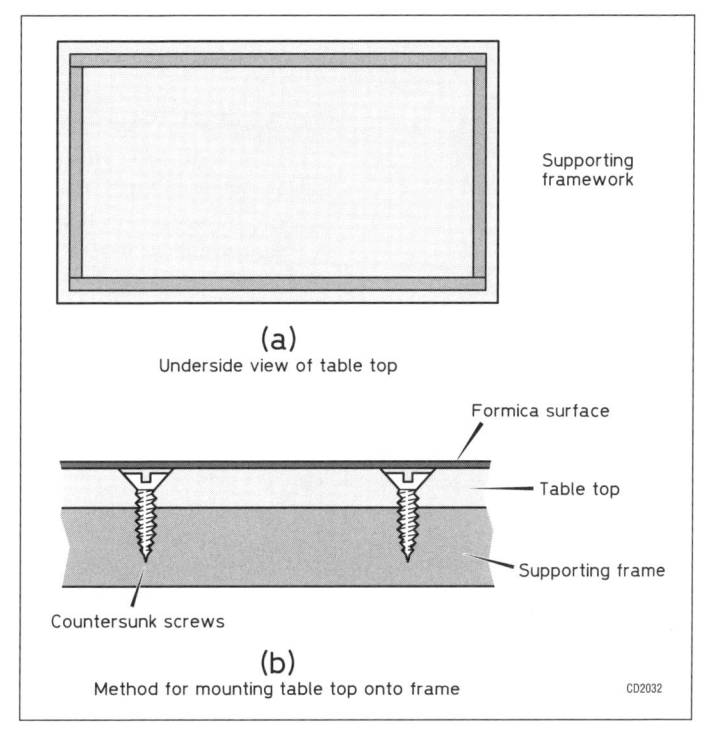

Fig 9.1. Construction of a table

Mains distribution

Some thought should be given to the way in which the mains power is distributed to the various pieces of equipment in the shack. The number of items requiring mains power can be surprisingly high and, if a single socket is used, this will quickly become overloaded. This problem can be overcome by using one or more of the mains distribution blocks containing four or five outlets in a straight line. These can be fastened to a suitable place near the back of the table, either on top or below the work surface. Then each piece of equipment can be permanently plugged in and turned on from its front-panel switch as required.

In addition to this the cables supplying these distribution blocks can be taken back to a common switch or circuit breaker so that the whole station can be turned off quickly and easily. It is also very wise to include a residual current circuit breaker. These items switch off the power when there is an imbalance in the levels of current taken between the live and neutral power lines, providing a very useful safety feature. They can be bought quite cheaply from most hardware or electrical suppliers.

A further item to consider is a single main switch to turn the whole station off. When the station is to be closed down this can be used to isolate all the equipment and ensure that nothing is left on accidentally.

Lighting

Lighting is an important feature in any shack, especially if any construction work is envisaged. One way of improving the lighting is to install a small strip lamp under a shelf above the table top to illuminate the work area. The front support on the shelf can then be used to shade the lamp from direct view. When deciding exactly where to fix it be careful to ensure that the manufacturer's recommendations for fitting and ventilation are obeyed.

An anglepoise-type lamp can be used if further lighting is required. These are ideal as they enable a large amount of light to be concentrated on the required area.

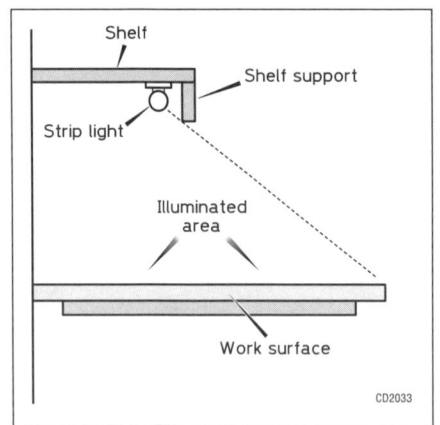

Fig 9.2. Lighting under a shelf can be used to illuminate the work area. Be careful to follow the manufacturer's instructions when fitting and do not mount it too close to the shelf support

Choosing equipment

There is an enormous choice of equipment available on the market. From small hand-held FM transceivers right up to the multi-mode, multi-band transceivers, there is something for almost every requirement. To choose the best buy for a particular station is not always easy. It is best to look carefully through the magazines for a while to see what is available from the dealers or second-hand in the readers' advertisements. Also, read the reviews to find out about the individual pieces of equipment.

It is also worth considering what is needed for the shack – the modes of operation envisaged, the power levels required, frequency coverage needed etc. In this way the field can be narrowed down.

It is sensible to look at some equipment before parting with any hard-earned cash. Mobile rallies, hamfests and the like are a very good opportunity for this. Many dealers come to these events and it is possible to look at a good number of units to see what they are like in real life rather than in pictures in magazines. It may also be worth a visit to a dealer – here it is often possible to sit down with a set, use it for a while and see how it 'feels' and whether it is the right one for you.

It is necessary to take extra care when buying second-hand equipment. It may have been in use for some time and might need attention – this is particularly true when buying privately. Most dealers have a name to protect and will not offer doubtful equipment for sale but this may not be the case with a private transaction. Check over the general appearance of the set – has it been well cared for or heavily used? First impressions count for a lot. If it has been heavily used, some components like switches may be worn and in need of replacement. Components may also have been replaced inside, and there may always be the possibility of further problems if the repair has not been completed properly. Also check to see if any modifications have been undertaken. If so, have they been implemented well? Often modifications are done by people who do not have sufficient skill

CHAPTER 9: SETTING UP THE RADIO STATION

Experimenter G4JNT has his test equipment neatly rack mounted. The use of a swivel chair ensures that everything is to hand

with a soldering iron, and they may cause more damage to the set than the modification is worth.

It is also necessary to check the general performance of the set to try to assess whether it works properly. Does it appear to be sensitive – can many stations be heard and at good strength? Are any beacons audible at about the right strength? Do the switches work properly, or are they intermittent if they are touched? Is the tuning smooth? Is the power output correct, and if a contact is made using a transmitter or transceiver, do the other stations report a good-quality signal?

It is often worth taking a friend along to look at a piece of equipment, especially if it is a private sale. A second opinion is always valuable and, if they have been in the hobby for some time and have experience of buying equipment, it can be helpful to draw on this.

While it is necessary to be careful about buying equipment, most people are honest and there are very many good deals to be found. By taking a little time to look through all the advertisements it is possible to find the best buy, whether new or second hand. When it has been bought and installed it will bring many hours of pleasure for your favourite aspect of amateur radio, or for trying out new areas of it.

Equipment layout

While ideas for the design of the shack and table are being formulated it is well worth giving some thought to the layout of the equipment itself so that the whole station comes together properly.

The main requirement is that the items that are used most should be conveniently placed so that they can be reached with the minimum of arm

G0MYX stacks his gear three shelves high. His work bench can just be seen on the right at 90° to the operating position

ache, especially when they are used for long periods, for example during a contest.

It is important that the tuning knob on the transceiver is correctly placed. This should be a couple of inches above the table surface and in a position where it can be reached easily. The microphone and Morse key should also be placed where they are easy to use. Often a microphone will be held with the left hand (for those who are right handed), leaving the right hand free to write notes. Similarly the Morse key, if one is used, will be operated using the right hand, and so it should be placed on the right-hand side of the table. It is also convenient to have space for a note pad and the logbook. Computers are often used in stations today for a variety of purposes from logging to connecting to the DX Packet Cluster or for propagation prediction software. Consideration should be given to locating the keyboard and screen so that they can also be used easily.

A typical station layout may have the main transceiver in the middle at the front with a linear amplifier to one side. A second receiver may be set to the other side. Ancillary equipment such as the ATU and VSWR bridge can be placed on a shelf over the transceiver and other main units. In this way they are easy to adjust and an eye can be kept on the meter readings. Space for a computer should be allocated if one is to be used. This might be set to one side, or the keyboard may be required on the main work surface, although it should not impede the operation of the main transceiver.

However, no two people's needs are the same and so each station will need to be planned on its own merits. To gain a few extra ideas it is worth looking at photographs of other stations that appear in the magazines quite regularly. Stations of the top DXers are particularly useful because they will

The neatly laid out station of GU4YOX enables him to use the Morse key and computer keyboard simultaneously

have spent many hours in their shacks using the equipment. On the way they are likely to have discovered many of the pitfalls of poor layout and reorganised them to be very easy and comfortable to operate. Finally, it is well worth investing in a good chair – it is no use trying to relax in the shack, or spend several hours operating in a contest, if the chair is uncomfortable.

Decorating the shack

Once the equipment in the shack has been set up the walls can be decorated. Maps like great circle maps, QRA locator maps, or prefix maps are very useful and give a lot of visual information very quickly. These should be easy to see, and may be on the wall behind the equipment.

It is also nice to put up some QSL cards, particularly those showing the best DX that has been worked. Unfortunately, mounting them on the wall can sometimes damage them. Pinning them to the wall obviously puts a hole in them, and Blu-tack can leave a mark after a while.

To overcome this problem it is possible first to mount the cards onto a postcard or some other suitable card using photograph corners. Then the postcard can be pinned or stuck to the wall, leaving the QSL cards free from damage. Alternatively transparent 'wallets' are obtainable from some photography suppliers.

General safety

It is probably true to say that safety standards in shacks have improved over the past few years. This is partly as a result of an increased awareness of the hazards, and partly due to the voltages in equipment being lower. Another reason is that more commercially made equipment is being bought which has to comply with certain safety standards before it can be imported

Newport Amateur Radio Society's station can be folded away into a cupboard when not in use

or sold. However, there are still a number of safety precautions which can be easily incorporated into the shack.

Electrical safety is of paramount importance, so ensure that all equipment is properly earthed. Sometimes there is a tendency to leave the earth connection off some pieces of equipment, which can be dangerous because it means that the whole of the case can rise to mains potential under certain fault conditions.

Earth leakage or residual current circuit breakers have already been mentioned, along with a single off switch for the whole station. These are well worth fitting.

While on the subject of electrical safety it is worth pointing out that all equipment carrying hazardous voltages or high levels of RF should be enclosed in cabinets. This is particularly important if visitors are likely to come into the shack at any time.

Finally, radiated RF should be kept away from inhabited areas of the house. Although it is unlikely that low power and non-directional antennas could cause any harm, this may not be true where high power and directional antennas are used. However, as it is difficult to assess field strengths at a particular place it is best to keep all RF at a distance.

These ideas represent only a few of the ways for keeping a shack safe. There are many other points that can make it safer. A general awareness of the dangers that might arise is the best solution. Then the shack will be a safer place for you and any visitors who may call in.

Further reading

Amateur Radio Operating Manual, 5th edn, ed Ray Eckersley, G4FTJ, RSGB, 2000.

APPENDIX

Abbreviations and codes

In this appendix:

- Commonly used abbreviations
- The Q code
- Phonetic alphabet
- RST code

Table A.1. Commonly used abbreviations	
ABT	about
AGN	again
AM	amplitude modulation
ANT	antenna
BCI	broadcast interference
BCNU	be seeing you
BFO	beat frequency oscillator
BK	break
B4	before
CFM	confirm
CLD	called
CIO	carrier insertion oscillator
CONDX	condition
CPI	copy
CQ	a general call
CU	see you
CUAGN	see you again
CUD	could
CW	carrier wave (often used to indicate a Morse signal)
DE	from
DX	long distance
ERE	here
ES	and
FB	fine business
FER	for
FM	frequency modulation
FONE	telephony
GA	good afternoon
GB	goodbye
GD	good
GE	good evening
GM	good morning

Table A.1. Commonly used abbreviations *(continued)*

GN	goodnight
GND	ground
HBREW	home brew
HI	laughter
HPE	hope
HR	here
HV	have
HW	how
LID	poor operator
LW	long wire
MOD	modulation
ND	nothing doing
NW	now
OB	old boy
OM	old man
OP	operator
OT	old timer
PA	power amplifier
PSE	please
R	roger (OK)
RCVD	received
RX	receiver
RTTY	radio teletype
SA	say
SED	said
SIGS	signals
SRI	sorry
SSB	single sideband
STN	station
SWL	short wave listener
TKS	thanks
TNX	thanks
TU	thank you
TVI	television interference
TX	transmitter
U	you
UR	your, you are
VY	very
WID	with
WKD	worked
WUD	would
WX	weather
XMTR	transmitter
XTAL	crystal
XYL	wife
Z	GMT – the letter is added after the figures, eg '1600Z' is 1600 hrs GMT
YL	young lady
73	best regards
88	love and kisses

APPENDIX: ABBREVIATIONS AND CODES

Table A.2. The Q code

QRA	What is the name of your station? The name of my station is ...	
QRG	What is my frequency? Your exact frequency is ...	
QRL	Are you busy? I am busy.	
QRM	Is there any (man made) interference? There is (man made) interference.	
QRN	Is there any atmospheric noise? There is atmospheric noise.	
QRO	Shall I increase my power? Increase power.	
QRP	Shall I reduce power? Reduce power	
QRQ	Shall I send faster? Send faster.	
QRS	Shall I send more slowly? Send more slowly.	
QRT	Shall I stop sending? Stop sending.	
QRU	Do you have any messages for me? I have nothing for you.	
QRV	Are you ready to receive? I am ready.	
QRZ	Who is calling me? You are being called by ...	
QSL	Can you acknowledge receipt? I acknowledge receipt.	
QSP	Can you relay a message? I can relay a message.	
QSY	Shall I change to another frequency? Change to another frequency.	
QTH	What is your location? My location is ...	
QTR	What is the exact time? The exact time is ...	

Table A.3. Phonetic alphabet

A	Alpha	N	November
B	Bravo	O	Oscar
C	Charlie	P	Papa
D	Delta	Q	Quebec
E	Echo	R	Romeo
F	Foxtrot	S	Sierra
G	Golf	T	Tango
H	Hotel	U	Uniform
I	India	V	Victor
J	Juliet	W	Whisky
K	Kilo	X	X-ray
L	Lima	Y	Yankee
M	Mike	Z	Zulu

Table A.4. RST code

Readability
1. Unreadable
2. Barely readable
3. Readable with difficulty
4. Readable with little difficulty
5. Totally readable

Strength
1. Faint, barely perceptible
2. Very weak
3. Weak
4. Fair
5. Fairly good
6. Good
7. Moderately strong
8. Strong
9. Very strong

Tone
1. Extremely rough note
2. Very rough note
3. Rough note
4. Rather rough note
5. Modulated note
6. Near DC note but with smooth ripple
8. Good DC note with a trace of ripple
9. Pure DC note

Index

A
A index 14, 16
Abbreviations, commonly
    used 115
AFSK (audio frequency
    shift keying) 25
AGC (automatic gain
    control) 41
ALC (automatic level
    control) 61
Alphabet, phonetic 118
Amateur bands 82
Amplitude modulation 21, 22
AMTOR 27
Angle of radiation 12
Antenna tuning unit
    (ATU) 66
Antennas 63
Ap index 16
Atmosphere, structure of .. 8
ATU (antenna tuning
    unit) 66
Audio filters 40
Audio frequency shift
    keying (AFSK) 25
Automatic gain control
    (AGC) 41
Automatic level control
    (ALC) 61
Automatic request for
    repeat (ARQ) 27
Awards 99

B
Balun 71
Band plans 81
Bands, amateur 81
    UK HF band list 5
    10m band 88
    12m band 87
    15m band 87
    160m band 82
    17m band 87
    20m band 85
    30m band 84
    40m band 84
    80m band 83
Baudot code 26
Beam antennas 67, 76
Beat frequency oscillator
    (BFO) 20, 37
Binary phase shift keying
    (BPSK) 29
Blocking, receiver 44
Break-in 53

C
Capture effect 24
Carrier 19
Characteristic impedance 65
Clipping, speech 55
Coaxial feeder 64
Codes
    Baudot 26
    Morse 19
    Q 117
    RST 118
Compression, speech ... 54
Contacts, basic 91
Contests 1, 99
Counterpoises 78
CQ call 92
Critical frequency 13
Cross-modulation 44
Crystal filter 40
CW 19

D
D layer 9
dBd unit 67
dBi unit 67
dBW unit 58
Dead zone 12
Decorating the shack ... 113
Deviation, FM
    transmission 24
Digital signal processing . 47
Dipole antenna 69
Direct-conversion receiver 34
Directivity of antennas ... 67
Doublet antenna 74
DX Century Club Award 99
DX nets 95
DX Packet Cluster 94
DXing 1, 93
DXpeditions 2, 94
Dynamic range 45

E
E layer 10
Earth system, antenna ... 78
Emergency
    communications 5
End effect in antennas ... 70
End-fed wire antenna 68
End-fire antenna 69
Equipment
    choosing 110
    layout 111

F
F layer 10
Feeders 64, 106
Fifth-order product 43
Filter, crystal 40
Flare, solar 14
Flux, solar 16
Frequency modulation ... 23
Frequency shift keying ... 24
Frequency synthesisers .. 37
Front-to-back ratio,
    antenna 68

G
Gain, antenna 67
Geomagnetic storm 14

119

Grey-line propagation ... 14
Ground system, antenna . 78
Ground waves 7

H
Harmonics, transmitter .. 59

I
Iambic-mode keyer 53
Image response 41
Incremental receiver
 tuning (IRT) 60
Intermediate frequency .. 36
Intermodulation
 receiver 43
 transmitter 59
Internet 94
Inverted-V antenna 72
Ionisation 9
Ionosphere 8, 9
Ionospheric sounding 13
Ionospheric storm 14
IRT (incremental receiver
 tuning) 60
Isotropic source 67

K
K index 14, 16
Key, Morse 21
Keying, transmitter 53

L
Lighting, shack 110
Linear amplifiers 57
Location, antenna 79
Logging programs 96
Lowest usable frequency
 (LUF) 13

M
Mains power in shack .. 109
Maximum usable frequency
 (MUF) 13
Mesosphere 8
Minimum discernible signal
 (MDS) 45
Mixing 34
Modem 25
Modulation 19
Morse code 19
Multiband dipole 73
Murray code 26

N
Narrow-band frequency
 modulation (NBFM) 24
Noise performance,
 receiver 41

O
Open-wire feeder 64

Operating techniques 91
Operating table for
 station 108

P
Pass band 39
Peak envelope power
 (PEP) 58
Phase noise 45
Phase-locked loop (PLL)
 synthesiser 38
Phonetic alphabet 118
Piezo-electric effect 40
Pile-ups 95
Polar cap absorption 14
Polar diagram 67
Poles, filter 40
Power
 shack mains 109
 transmitter 58
Predictions,
 propagation ... 15, 94
Pressel switch 52
Product detector 23
Propagation, radio wave .. 7
PSK31 28

Q
Q code 117
QRP
 operation 20, 89
 transmitters 49
QSL cards 2, 97
Quadrature phase shift
 keying (QPSK) 29

R
Radials 78
Radio teletype (RTTY) .. 25
Radio wave propagation .. 7
Receivers 33
Reciprocal mixing 45
Reflections, radio wave .. 10
Repeaters 24, 89
Ribbon feeder 64
Rotator, antenna 68
RST code 118
RTTY (radio teletype) ... 25

S
Safety
 antenna 80
 shack 113
Selectivity 39
Sensitivity 41
Shack, radio 105
Shape factor 40
Short-wave bands 5
Sidebands 22
Single sideband (SSB)
 receivers 22
 transmitters 51

Skip distance 12
Skip zone 12
Sky-wave 8
Sloper antenna 72
Slow-scan television
 (SSTV) 29
Solar flare 14
Solar flux 16
Speech processing 53
Split frequencies 95
Spurious signals 58
SSB (single sideband) ... 22
SSTV (slow-scan
 television) 29
Standing wave ratio
 (SWR) 66
Station, radio 105
Stop band 39
Stratosphere 8
Strong-signal handling
 performance 42
Sudden ionospheric
 disturbance (SID) .. 14
Sunspots 11
Superhet receiver 36
SWR (standing wave
 ratio) 66

T
Terminal node controller
 (TNC) 25
Thermosphere 8
Third-order intercept 43
Third-order product 43
Top Band 82
Transceivers 58
Transmission lines 64
Transmitter power 58
Transmitters 49
Trap dipole antenna 74
Trap vertical antenna ... 76
Troposphere 8
Twin feeder 64

V
Varicode 29
Velocity factor 65
Vertical antenna 75
VFOs, separate 60
VOGAD (voice operated
 gain adjusted device) 54
VOX 52
VSWR tolerance,
 transmitter 60

W
Wide-band frequency
 modulation (WBFM) 24

Y
Yagi antenna 77

MORE BOOKS FROM THE RSGB

HF Antenna Collection

Edited by Erwin David, G4LQI

An invaluable collection of the outstanding articles and short pieces that were published in the Radcom magazine during the period 1968-89. Includes ingenious designs for single element, beam and miniature antennas, as well providing comprehensive information about feeders, tuners, baluns, testing, modelling, and how to erect your antenna safely.

1st Edn, 1992, RSGB, paperback, 184 by 245 mm, 233 pages, ISBN: 1-872309-08-9.

Price: **£9.99**

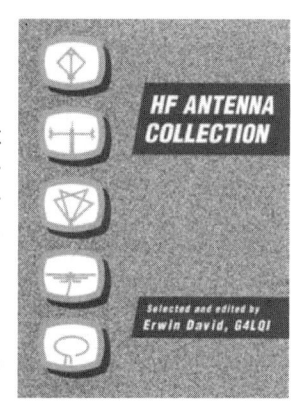

The Antenna File - NEW

The Radio Society of Great Britain produces some of the best works on antennas and this is a collection of that work from the last ten years. This book contains 288 pages of articles drawn from the Radcom magazine and includes: · 50 HF antennas, 14 VHF/UHF/SHF antennas, 3 receiving antennas, · 6 articles on masts and supports, · 9 articles on tuning and measuring. · 4 on antenna construction. · 5 on design and theory · And 9 Peter Hart antenna reviews. · Every band from 73kHz to 2.3GHz · Beams, wire antennas, verticals, loops, mobile whips and the G2AJV Toroid. In fact everything you need to know about antennas and how to get the best out of them.

1st Edn, 2001, RSGB, paperback, 297 by 210 mm, 288 pages, ISBN: 1-872309-72-0.

Price: **£18.99**

HF Antennas for all Locations

By Les Moxon, G6XN

This is a thought-provoking book, which has been a major contribution to the state of the art from an acknowledged expert. It explains the 'why' as well as the 'how' of HF antennas, and takes a critical look at existing designs in the light of the latest developments. This second edition has been completely revised and greatly expanded. There are more novel antenna designs, including beams which cover more bands with fewer problems, no trap losses and better rejection of interference. A new chapter presents a comprehensive review of ways to make antennas smaller, with particular emphasis on small transmitting loops. An essential reference for the experimenter and enthusiast.

2nd Edn, 1993, RSGB, paperback, 187 by 245 mm, 322 pages, ISBN: 1-872309-15-1.

Price: **£7.99**

www.rsgb.org/shop Tel: 0870 904 7373

RSGB ORDER FORM

ORDERED BY

ORDER NO.　　　　　**DATE**

DELIVER TO

Code	Description	Price	Qty	Total
1-872309-54-2	Backyard Antennas	£18.99		
1-872309-08-9	HF Antenna Collection	£9.99		
1-872309-72-0	The Antenna File **NEW**	£18.99		
1-872309-15-1	HF Antennas for all Locations	£7.99		
1-872309-11-9	Practical Antennas for Novices	£7.99		
1-872309-36-4	The Antenna Experimenter's Guide	£17.99		
1-872309-74-7	RSGB Yearbook 2002 **NEW (available Sept)**	£15.99		
1-872309-53-8	Radio Communication Handbook	£29.99		
1-872309-65-8	Low Frequency Experimenter's Handbook **NEW**	£18.99		
1-872309-40-2	PMR Conversion Handbook	£16.99		
1-872309-30-5	Radio Data Reference Book	£14.99		
1-872309-35-6	Practical Receivers for Beginners	£14.99		
1-872309-21-6	Practical Transmitters for Novices	£16.99		
1-872309-23-2	Test Equipment for the Radio Amateur	£12.99		
1-872309-61-3	Technical Topics Scrapbook 1995-99	£14.99		
1-872309-51-8	Technical Topics Scrapbook 1990-94	£13.99		
1-872309-20-8	Technical Topics Scrapbook 1985-89	£9.99		
1-872309-71-2	The RSGB Technical Compendium **NEW**	£17.99		
0-705652-1-44	Radio & Electronics Cookbook **NEW**	£16.99		
1-872309-73-9	Low Power Scrapbook **NEW**	£12.99		
1-872309-00-3	G-QRP Circuit Handbook	£9.99		
0-900612-89-4	Microwave Handbook **Volume 1**	£11.99		
1-872309-01-1	Microwave Handbook **Volume 2**	£18.99		
1-872309-12-7	Microwave Handbook **Volume 3**	£18.99		
1-872309-48-8	The RSGB Guide to EMC	£19.99		
1-872309-58-5	Guide to VHF/UHF Amateur Radio **NEW**	£8.99		
1-872309-42-9	The VHF/UHF Handbook	£19.99		
1-872309-63-1	Amateur Radio Operating Manual **NEW**	£24.99		
N/A	Prefix Guide (fifth edition, 1999)	£8.99		
1-872309-43-7	Your First Amateur Station	£7.99		
1-872309-62-3	The RSGB IOTA Directory	£9.99		
1-872309-31-3	Packet Radio Primer	£9.99		
1-872309-38-0	Your First Packet Station	£7.99		
1-872309-49-6	Your Guide to Propagation	£9.99		
1-872309-60-7	Radio Today – Ultimate Scanning Guide	£19.99		
1-872309-27-5	Novice Licence - Student's Notebook	£4.99		
1-872309-28-3	Novice Licence - Manual For Instructors	£9.99		
1-872309-45-3	Radio Amateur's Examination Manual	£14.99		
1-872309-18-6	RAE Revision Notes	£5.00		
1-872309-19-4	Revision Questions for the Novice RAE	£5.99		
1-872309-26-7	Morse Code for Radio Amateurs	£4.99		
1-872309-50-0	Amateur Radio ~ the first 100 years	£49.99		
0-900612-09-6	World at Their Fingertips	£9.99		

Post & Packing — P&P

UK only - £1.50 for 1 item £2.95 for 2 or more items — Discount

Rest of World - £2.00 for 1 item 4.00 for 2 & £0.50 for each extra item — Total

RSGB, Lambda House, Cranborne Road, Potters Bar, Herts EN6 3JE UK
Tel: 0870 904 7373 Fax: 0870 904 7374 E-mail sales@rsgb.org.uk